D0007482

WHAT YOUR CAT KNOWS

WHAT YOUR CAT KNOWS

Tap into your cat's intelligence through
the world of feline cognition

SALLY MORGAN

METRO BOOKS
NEW YORK

An Imprint of Sterling Publishing Co., Inc.
1166 Avenue of the Americas
New York, NY 10036

ISBN 978-1-4351-6565-6

For information about custom editions, special sales,
and premium and corporate purchases,
please contact Sterling Special Sales at 800-805-5489
or specialsales@sterlingpublishing.com.

Manufactured in China

4 6 8 10 9 7 5 3

www.sterlingpublishing.com

Credits: Design and illustrations by Matt Windsor

"Of all God's creatures, there is only one that cannot be made slave of the leash. That one is the cat."

—**Mark Twain**

CONTENTS

INTRODUCTION

What is your cat thinking as she eyes you with her inscrutable stare? What is she plotting? Is she happy? Does she even like you? It's clear that cats are an enigma. Their lack of facial expressions mean it can be difficult to figure out what is going on in the minds of cats. There's no wagging tail or excited bark that we get with dogs.

A SPECIES APART

Cats are more of a blank page, with few facial grimaces to help us. They are aloof, fiercely independent animals with a curious streak. They can be playful as well as stand-offish. They communicate with us, but on their terms and often they make it clear that they don't want our company.

The history of the cat is different to that of the dog. Cats have lived around us for almost as long as dogs, but because they have never been totally dependent upon us for food or shelter they have retained their independence. We put them to work as pest controllers in grain stores and they earned their keep by catching rats and mice. Consequently, they have retained their wild side, and today it bubbles just below the surface. During the day, your average cat stays close to home, eating, sleeping, and playing; but as dusk falls their wild instincts take over and they like nothing more than to head outside and hunt for prey. And most of us don't have a clue about what our cats get up to while they are patrolling the streets at night!

A cat's world is very different from our own. Their incredible sense of smell, for example, enables them to detect many more scents than we can. Many owners even claim that their cat is psychic, and it seems some cats can detect sickness in their owner before any diagnosis by a doctor.

We remain in the dark about cat relationships, too. How many cat owners can honestly claim to understand the relationships between their cats, or even if their cat is happy or stressed? There are so many feline signs that we miss and, as a consequence, our relationships with cats falter.

WILDCATS

In genetic terms, cats are not that different from their wildcat ancestors. Analysis of DNA has shown that dogs and wolves have diverged far more than domestic cats have from wildcats. That's not really surprising, as people have been selecting and breeding dogs for hundreds of years, choosing particular appearances and characteristics, while cats have undergone far less selective breeding. It's only in the last 150 years or so that pedigree breeds of cat have been recognized. However, in recent years, the so-called hybrid cats have become popular. These are breeds that have been produced by breeding back to species of wildcat. They include the Bengal (African leopard cat), Savannah (African serval), Chausie (jungle cat) and, most recently, the Marguerite. The Marguerite is a result of a cross between a captive-bred African sand cat and a British crossbreed.

Until recently there had been little research into the mind of the cat, but, as researchers unlocked some of the secrets of the dog, they have begun to turn their attention to the cat. However, cats are not the easiest of animals to work with. In fact, one researcher claims it is easier to work with fish! Despite this, research is beginning to reveal some amazing insights into their psyche, and this can only improve our relationships with cats.

SECTION ONE
THE FELINE SENSES

We think we know our cats and their world, but honestly, we know nothing. You may not realize it, but you are living with one super-sense feline friend.

Just like us, cats have five senses—sight, hearing, smell, taste, and touch—but that's where the similarity ends. If it were a competition, our cats would beat us every time.

Their super-senses explain a lot: how they know that we have changed their brand of food or are using a different type of litter; how they can run around the backyard in the dark without injuring themselves; and how they fall off a wall and land on their feet. And then there is the extra-sensory stuff—how do they know a storm is on its way, an earthquake is about to hit, or even that somebody is about to die?

First, let's find out about their senses and learn about the world from a cat's point of view.

CHAPTER 1: SUPER-VISION

Who could fail to notice the beautiful cat's eye, with its distinctive vertical slit? But it doesn't just look good; its design is perfect for a nighttime hunter. And, as we will discover, the shape of the pupil is just one adaptation that gives the cat super-vision.

DAWN AND DUSK

As my cat slips outside at night, I never fail to wonder how he manages to find his way around in the dark. I can't see much, even when I stand outside to let my eyes get used to the dark, but my cat can easily navigate around the farm and fields and catch mice.

Our pet cats have descended from wildcats that lived in the Middle East, where the high daytime temperatures meant that they rested during the day and emerged at dusk, when it was much cooler. Wildcats are most active at dawn and dusk. Despite thousands of years of domesticated life, our cats have retained this pattern of behavior, which is why they just love to venture out at night to hunt.

SIZE MATTERS

One adaptation that has resulted from this lifestyle is large eyes. In fact, your pet cat's eyes are the same size as your own, but in a much smaller head, so they are proportionately much larger. Interestingly, cats' daytime sight is pretty poor and we can see far more detail, but our cats win hands down at night. So how do they get such great nighttime vision? It's all to do with an abundance of light-sensitive cells and a special mirrorlike layer at the back of their eye.

The eye works by focusing the light reflected off objects in the field of view to the back of the eye, where it is received by light-sensitive cells. First the light passes through the cornea to the lens, after which the lens bends the light, focusing it to a point on the retina. Light-sensitive cells in the retina, called rods and cones, detect the light and send a message to the brain along the optic nerve.

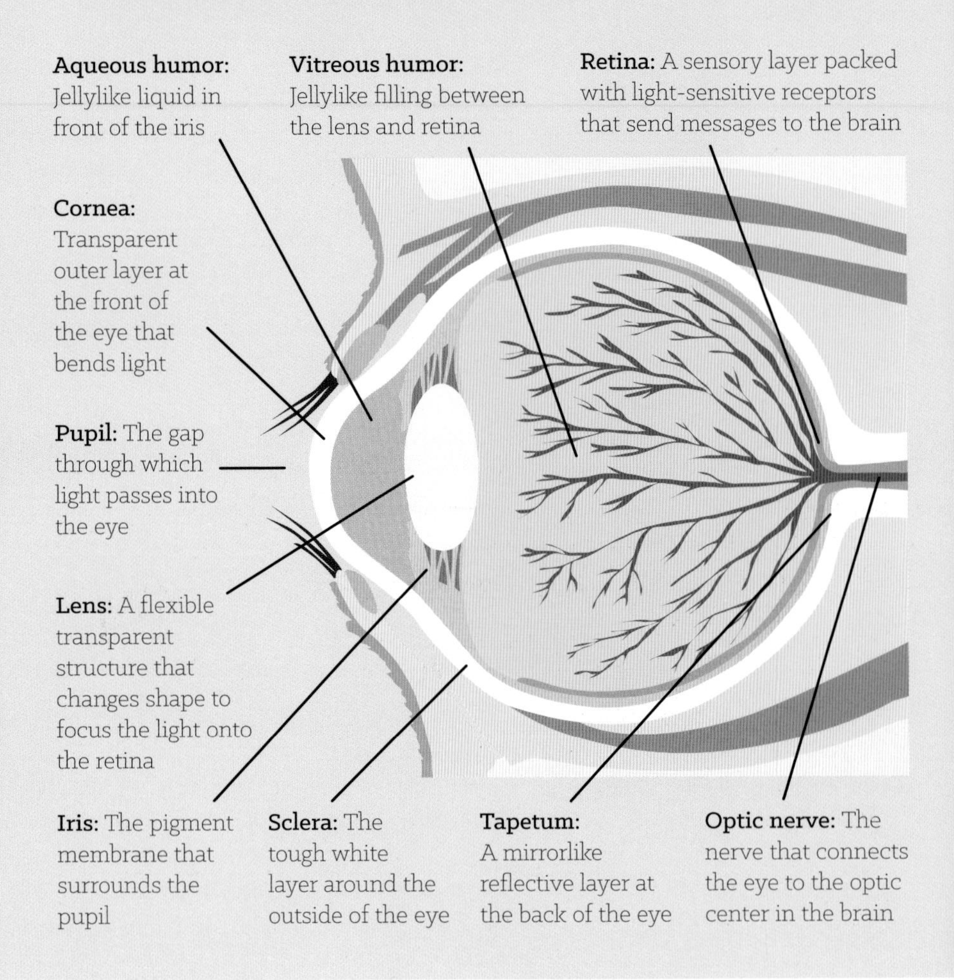

RODS AND CONES

The retina is found at the back of the eye and is packed with two types of light-sensitive cells—rods and cones. The key to a cat's super night vision is the large number of rods in a cat's eye, which are responsible for seeing in low-light levels. They enable the cat to see in the dark. However, the cones, which detect color, only work in bright light. Cats have about six times more rods than humans, and this enables them to see much better in very poor light. Cat's eyes are also better at picking up movement in the dark, spotting the sudden dash of a mouse or rat. Overall, it's estimated that cats can see eight times better than us in the dark.

Our own retina is cone-rich, and there is even one point on the retina, called the fovea, that consists only of cones and it is responsible for extra sharp vision. Cats lack a fovea; they instead have a visual streak, an area that has a high concentration of rods. It helps them to track movement. However, without the fovea, they can't see quite as clearly as us.

Both cats and humans have fewer cones in the periphery of the eye, which means that we both have to move our eyes or head to keep a sharp focus on a moving object.

DID YOU KNOW?

Your cat's eyes, or more specifically the size of the pupils, tell you a lot about her state of mind. A happy, relaxed cat has the usual vertical slit. However, if the pupils become wider in bright light, there is something else going on. Pupil size is linked to the "fight or flight" response; if a cat feels threatened or fearful, her pupils get larger to allow in more light. This helps the cat to see more clearly and prepares her for a possible escape. A vet soon learns to look out for a cat with dilated pupils, as she may be in pain, about to jump, or perhaps ready to bite or scratch.

VERTICAL PUPIL

We have a round pupil, as do the big cats, such as lions and tigers. However, domestic cats, snakes, and crocodiles have a vertical pupil, which is key to seeing in the dark. In bright light, the pupil is a narrow, vertical slit, whereas in low light it opens wide, becoming almost round. The huge difference between fully constricted and fully dilated equates to a roughly 300-fold increase in the area of the pupil, while our round pupils can achieve only a fifteen-fold increase. This gives cats a huge advantage when hunting, as their pupils can open really wide to let in the maximum amount of light.

In bright light, our pupils shrink until they are just a pin prick, stopping too much light from entering the eye and damaging the retina. Being adapted to seeing well in low light, you would think that a cat's eye would be easily dazzled in bright light, but this is where the cunning design of the vertical pupil comes into play. The shape is great for letting in a lot of light in the dark, but also helps to cut down the light entering in bright conditions. As the level of light increases, the pupil narrows to a tiny slit. Then, to reduce the light further, the cat half closes her eye lids and squints. It's a perfect design.

The vertical pupil is also thought to give cats an edge when hunting, as it allows them to get a better estimate of their prey's distance and to focus more accurately on their target.

GLOW IN THE DARK

Have you ever noticed that your cat's eyes have a slight green glow when they catch a source of light in the dark, such as from a camera flash? The same is true of other animals, such as dogs and deer, but not of our own eyes. This weird effect is due to a special reflective layer that lies behind the retina in the back of the eye. Called the tapetum, it is packed with mirrorlike cells that reflect light. Any light passing through the retina and falling on the tapetum is reflected back into the eye, making sure no rays of light are wasted. This is important because, in darkness, every glimmer of light can be captured to help the cat see better.

There is also evidence that the wavelength of light reflected back off the tapetum is different from the wavelength falling on it, and that this may help to enhance the image that the cat sees. For example, it would make a shape stand out against the background. All of this means that cats are unparalleled nighttime hunters.

DID YOU KNOW?

Not all cat's eyes glow the same color. Most breeds have eyes that glow bright green, but Siamese cats have a bright yellow glow. This is due to the amount of pigment in the retina and the presence of substances such as zinc in the tapetum. A few cats, especially white cats with blue eyes, lack a tapetum altogether. And if you were wondering how the safety devices used on roads, which are commonly known as "cat's eyes," work, the inventor, Percy Shaw, based his design on the way a cat's eyes reflect light in the dark.

Underground
Way out
Standard Express
Customer lounge
Cycle store
First Class lounge
Left luggage

CAN YOUR CAT SEE COLOR?

Cats can't have it both ways. There is a downside to having great night vision: they can't see as well in daylight. As well as not seeing a lot of detail, cats can't make out many colors either. Because of this, during the day, their other senses are just as, or more, important than their sight.

Cones are responsible for color vision and accurate detail, but they need bright light to work well. This is why we can only see shades of black and white in low light. Our eyes have three types of cones, each sensitive to a different wavelength—namely, red, green, and blue, the primary colors. The three types of cones work together to detect all the colors of the visible spectrum. Because human eyes contain a lot of cones, the colors look vibrant to us.

Scientists are sure that cats have some color vision, because cats have been trained to distinguish between red and green, red and blue, and red and yellow lights. For this to be possible, cats must have at least two types of cones. Some scientists think that cats have three types of cones, like us, but in far fewer numbers, which means that they can't see the same range of colors as we can. It is most likely that they can detect shades of blue and yellow and a bit of green, along with gray, white, and black. Whatever the colors, they are certain to be less saturated, because they have so few cones. Most scientists agree that cats are unlikely to be able to detect the difference between hues of orange and red, and between red and green, so they probably see an image similar to that seen by a person who suffers from red/green color blindness. However, for a cat, color is not as important as the shape, size, and pattern of an object.

SEEING ULTRAVIOLET

How many times have you seen your cat fascinated by a simple piece of white paper? Well, it seems that the cat's not crazy, but may be seeing things that we cannot!

We have known for a long time that bees and other insects can see ultraviolet (UV) light reflected from flowers, which helps them find nectar. Rats can see UV light too, using it to follow urine trails along the ground. In the Arctic, reindeer use it to spot polar bears, as the bear's fur does not reflect UV but the snow does. Recent research, led by Professor Ronald Douglas at City University of London, has revealed that cats, too, can see UV light wavelengths.

Humans can't see UV light because the lenses in our eyes block it out. This is thought to be a protective mechanism; UV light tends to make things more blurry, so by filtering it out, our eyes retain their sharp vision. Skiers often wear yellow goggles to block out the UV light reflected from the snow, so that they can see where they are going more clearly.

In the home, many products use optical brighteners. These are substances that absorb UV light and glow, or fluoresce. These brighteners are used to enhance the appearance of products, such as paper, cosmetics, washing powder, shampoos, detergents, and fabrics, to give them a "whiter-than-white" appearance. With brighteners, a piece of white paper may look very different to a cat's UV-sensitive eyes. And just imagine what we look like to the cat, if we have used cosmetics and shampoos with optical brighteners!

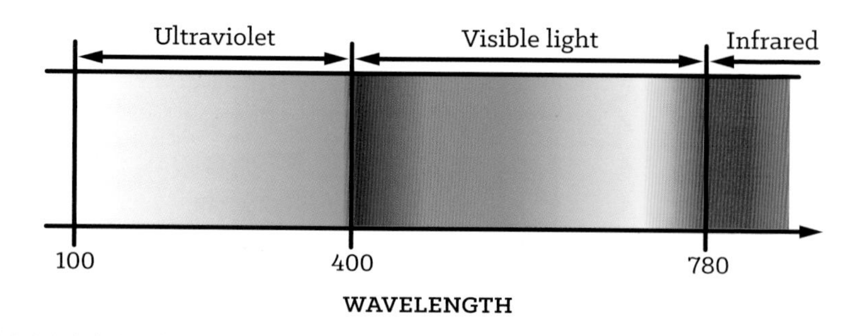

UP CLOSE

Our eyes can focus on objects that are very near as well as farther away because the lens in each eye changes shape. They can change from round and fat to long and thin, and this changes the curvature of the lens. This in turn alters the way the light rays are bent, ensuring that light from objects, no matter how distant, can be focused on the retina.

Cats can't do this because the lens in each eye has a fixed shape. This prevents cats from focusing on close-up objects, meaning they need to be farther away from an object to see it clearly. The nearest they can see clearly is 10 inches (25cm); any closer and it's a blur. It's a bit like trying to take a close-up photo of a flower—it's all blurred unless you use a special macro lens. So, if cats can't see clearly in front of their face, how do they manage when things are up close? They get around this by relying on their sense of smell and their sensitive whiskers, which can send information to the brain.

HOW SHARP IS THEIR VISION?

You may have heard your optician use the term "20/20 vision" when you have had your eyes tested, which refers to the ability to see an object 20 feet (6m) away in sharp focus. The two "20"s in the name refer to the distance in feet from the chart and the size of letters that can be seen clearly. In an eye test, you are asked to read the letters on an eye chart. The rows of letters get smaller and smaller and, if you have 20/20 vision, you can read letters at the bottom that represent "normal vision." If you can read the tiny letters on the lowest line, you have better than average vision (20/15 or 20/10, perhaps), but, if your eyes are worse than average, you will only be able to read the larger letters, and your vision may need to be corrected (20/100, for example).

Cats have worse vision than us, typically between 20/100 and 20/200, so they would only be able to see the large letter at the top of an eye chart clearly. In practical terms, if you stood beside your cat outside, he could see an object 20 feet (6m) away as well as you could see it from 100 feet (30m) away (assuming you have 20/20 vision). This probably means that when you are walking toward your cat, your face will look a bit blurry.

Acuity is the ability to see detail. Each of these drawings shows a series of lines. A human can see thirty lines per degree of the visual field; dogs can see twelve lines; but cats can see only six. Hence, they do not see as much detail as us.

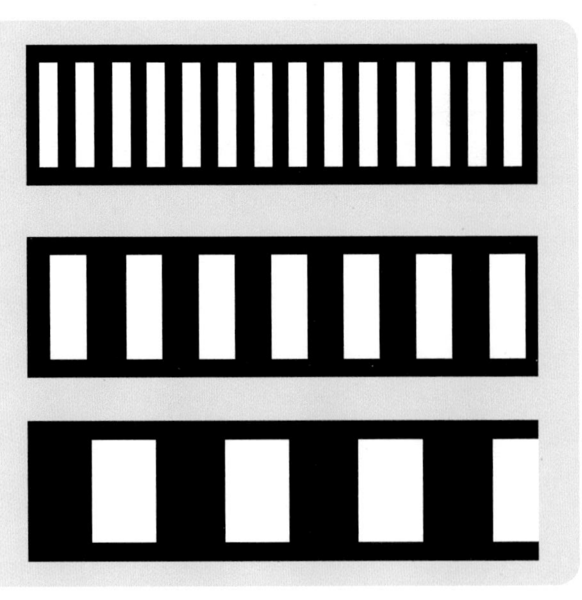

FIELD OF VIEW

If you look straight ahead and focus on a single point, all that you can see ahead, above, below, and to the side without moving your eyes is your field of view, or visual field. We can see an arc of about 180 degrees in front of our eyes, but a cat has a wider field of view of around 200 degrees. This means he is able to see more out the corner of his eye, such as a fleeing mouse or fluttering bird.

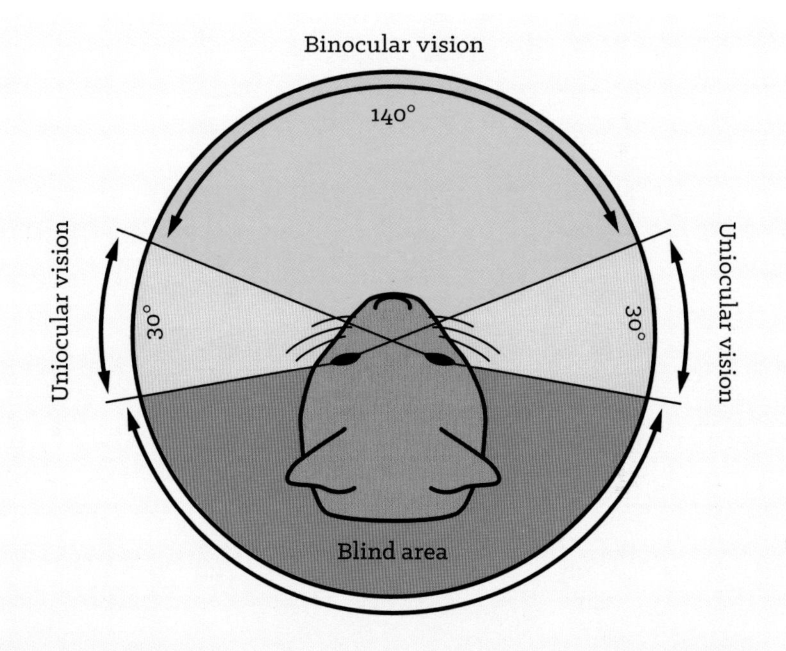

Try out this simple test on yourself. Hold a colored crayon in each of your hands, with your arms held out sideways. Keep your eyes looking forward, fixed on a detail ahead of you, and see if you can detect the crayon out the corner of your eyes, in your peripheral vision. Now try this on your cat. Wait until they are sitting and looking forward. Hold the crayons out to the side of their head and see if they spot you wiggling the crayon out of the corner of their eye.

3D VISION

Like all predatory mammals, cats have two forward-facing eyes that give them 3D, or binocular, vision, which is perfect for looking forward and downward and is ideal for hunting. Their brain compares the images from the right and left eyes to create a 3D image. This 3D picture enables them to judge distance and so pounce accurately on their prey or leap from roof to roof. The next time you see your cat out hunting, watch carefully as he prepares to pounce. Often you will see tiny side-to-side head movements, which is the cat's way of working out distances. The slight movement is enough to give very accurate estimates of the distance, which is critical, because the cat doesn't want to pounce and miss.

MOVEMENT DETECTORS

Whereas we can see slow-moving objects, cats cannot, although they will react to a fast-moving one. Their movement detectors enable them to respond to the tiniest of movements, such as the flicker of an ear or the bend of a blade of grass. Cats are not born with this ability; they have to learn it as kittens. Their brains learn how to process as many as sixty images per second, analyzing them for the tiniest change or movement that would give away the presence of a prey animal. This is an amazing skill that is difficult to comprehend.

WHAT DOES YOUR CAT SEE ON TV?

Cats may "watch" TV, but their experience is very different from our own. They don't see all the colors and can't see fine detail either, so the screen will look washed out, lack color, and appear a bit blurry. What's more, the moving image on the screen is made up of frames that are refreshed many times per second. Anything faster than twenty frames per second and we see a continuous film rather than a series of images. But cats can process visual information far faster, so they see a series of flashing images rather than a moving film. No wonder they don't show much interest in TV!

Funnily enough, this isn't true for all domestic cats. Some breeds, such as the Burmese, are not as good at visual processing and see our TV screens just fine. Birds, however, with an even greater ability to process visual information, can get very distressed in a room with a flickering TV.

A THIRD EYELID

Have you noticed that, when cats stare intently at a bird or a toy, they don't blink? That's because the tiniest movement of their eyelids could be enough to give away their presence to their prey and enable it to escape.

Blinking is important for moisturizing the surface of the eye, which is why humans blink very regularly. Cats blink too, but not nearly so frequently. Like many other mammals, cats get around this by having a third eyelid or, to give it its proper name, a nictitating membrane. It's a fold of tissue that faces the cornea on one side and the inside of the eyelid on the other.

When the cat is awake, most of his nictitating membrane is tucked away within the eye socket and only a tiny bit is visible in the inner corner of the eye. During sleep, or when he blinks, the membrane is wiped diagonally across the eye from the inner corner to the upper corner, just like a windshield wiper, removing dirt and debris from the surface of the cornea and rinsing it with tears. This helps to keep the surface of the eye healthy and moist.

It probably protects the eye, too, especially when the cat is running through long grass, trying to hang on to a wriggling animal armed with a sharp beak or claws, or having a fight with another cat.

DID YOU KNOW?

Blinking or fluttering their eyelashes is a sign of affection and trust in the cat world.

A BLIND CAT

Cats can become blind for many reasons—they may have suffered a trauma, an illness, or one of several progressive diseases that affect older cats. But cats are remarkable animals that are able to compensate for and adapt to the complete or partial loss of their sight. As we know, their daytime vision is not great and they already rely on their other senses, especially smell and touch, so it's not surprising that they can adapt well. In fact, their whiskers can become substitute eyes, allowing them to feel their way around.

What a blind cat needs is a consistent home environment, so the furniture shouldn't be moved around too much. Their food and water should be put down in the same place and there shouldn't be much clutter lying around. Since they can't go outside, a blind cat will benefit from an enriched environment with lots of stimulation of their other senses—toys with rattles and bells, smelly things, and even a path of catnip around the home.

CHAPTER 2: ALL ABOUT THE CAT'S EAR

Have you noticed that your cat can swivel his ear flaps to listen to what is going on behind him without moving his head? And he can move each ear independently, too. Freaky! That's thanks to the 32 muscles in each of his ears that allow him to swivel and twist the ear by as much as 180 degrees to find the source of a sound and then amplify it. In comparison, we have a mere six muscles and, if we're very lucky, we may be able to wiggle our ears slightly. A cat's exceptional hearing means he usually knows when somebody is walking up to the front door long before you can hear anything.

GOOD RANGE

Cats have the broadest hearing range in the mammalian world. Most mammals have a narrow hearing range, hearing either high or low frequency sounds, but cats are different. Not only are they able to hear the squeaks of mice and other high-frequency sounds, but they have the ability to hear low-frequency sounds, too. This gives them a hearing range from 48 hertz (Hz) to as high as 85,000 Hz, if the sound is loud enough. This compares with our range of around 20–20,000 Hz. That means that cats have a range of eleven octaves and can hear two octaves higher than us, opening up a whole world of sound that we are oblivious to.

DIRECTION FINDERS

The 32 muscles in each of the ear flaps enable a cat to swivel his ears independently to locate the source of a sound. Cats are able to locate the source of a sound with great accuracy; it has been claimed that they can distinguish between two sounds originating barely 4 inches (10cm) apart yet 40 inches (100cm) away. Their brains do all the processing of the information, and are aided by the fact that the ear nearest to the sound hears the sounds first—not by much, but it's enough to enable their brains to analyze the sounds coming from the right and left ear separately and then compare the results. This means that a cat can use all this sound information to build up a 3D "sound picture," enabling him, for example, to locate the position of a mouse in the undergrowth and pounce with astonishing accuracy. For wild cats, this ability is a matter of survival.

DID YOU KNOW?

A cat's ear is a bit of a "mood barometer," and it can tell us about their emotions. An annoyed cat lowers his ears (see photograph, below left); a terrified cat flattens his ears against the side of his head (below right); and an inquisitive cat has his ears well forward (bottom).

ALL ABOUT SOUND

Sound is a type of energy caused by particles vibrating back and forth and bumping into each other. This movement, known as a pressure wave, spreads out rather like a ripple spreading out across a pond until it runs out of energy. Sound travels at 1,125 feet (343m) per second in air and 4,869 feet (1,484m) per second in water.

The distance between the waves is called a wavelength. A sound's frequency is the number of waves per second that pass a fixed point, and it is measured in hertz (Hz). For example, a sound of 20 Hz has twenty waves per second passing a fixed point. A high-pitched sound, such as the squeak of a bat, will have waves close together, whereas a low-pitched sound, such as the call of a whale, has widely spaced waves. The loudness of a sound is determined by the size, or amplitude, of the sound wave. A wave with greater amplitude has more energy and will therefore be louder.

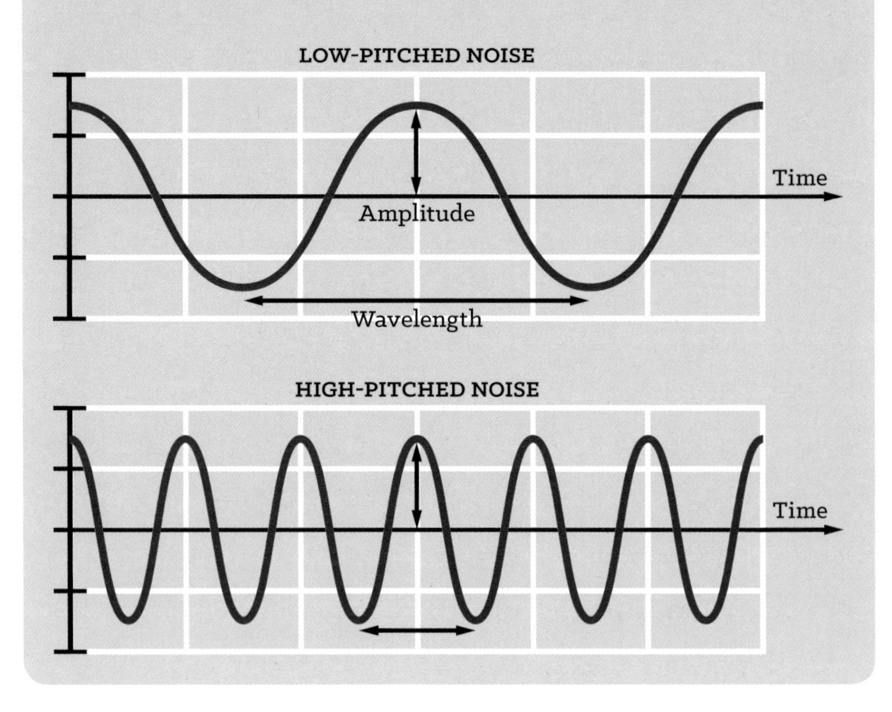

ULTRASOUND

A cat hears far more than we do. The ultrasonic squeaks of mice, rats, voles, and bats are very high-pitched and are well above our hearing range, so we are completely unaware of them, unless we use a special detector. Kittens can produce very high-pitched meows when they get separated, which helps to attract their mother.

At the other end of the scale are infrasonics, sounds that are below our hearing range. Infrasound is important in nature. For example, part of the roar of the lion and tiger is at a booming low frequency that we can't hear, but other big cats can. Infrasound travels over long distances and helps the lion and tiger to maintain a large territory. The deep rumbles of earthquakes and avalanches produce infrasound, which is why cats often react when we fail to notice anything at all.

DID YOU KNOW?

Many of the deterrents used to keep unwanted cats out of backyards make use of the cat's ability to hear ultrasounds. The high-pitched sounds that the deterrents produce can be heard by the cat but not the neighbors.

HOW DOES THE EAR WORK?

The ear comprises three parts: the outer, middle, and inner ear. The outer ear collects and directs sound into the middle ear, where it is magnified before being passed to the inner ear, where it is detected by sensory receptors and information is sent to the brain.

The little tufts of hair in a cat's ear direct sounds into the ear, help keep out dirt, and insulate the ears. They are called "ear furnishings." The pinna, or ear flap, channels the sounds down the ear canal to the eardrum, causing the eardrum to vibrate. These vibrations push on the first of three small bones (the ossicles) lying in the middle ear. The first bone vibrates and pushes on the second, which in turn pushes on the third, which touches the oval window—an opening that leads to the inner ear. The vibrations of the three ossicles magnify the sound by roughly 22 times.

The inner ear comprises a long, coiled fluid-filled tube called the cochlea. When the membrane across the oval window vibrates, it pushes on the fluid, causing that to vibrate, too. The fluid pushes on sensory receptors, which send messages along the auditory nerves to the brain.

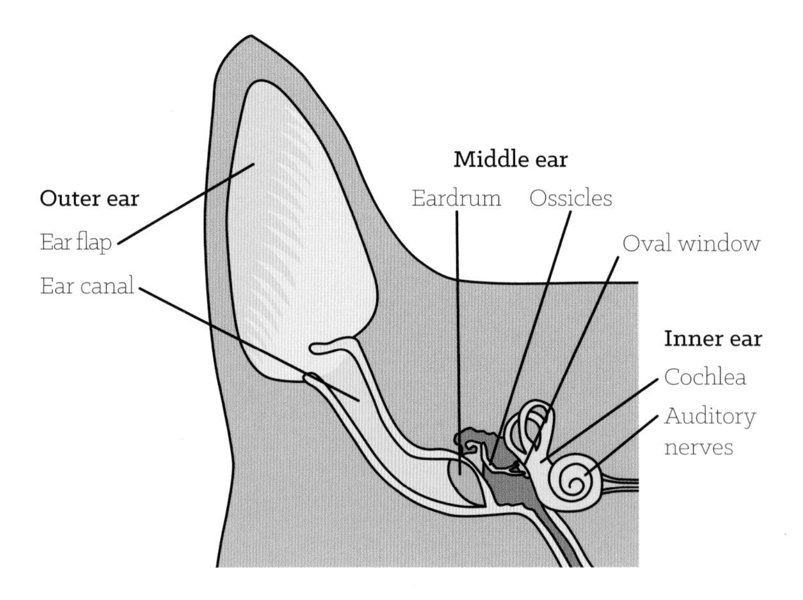

ODDLY SHAPED EARS

Most cats have upright ears supported by a piece of cartilage, which is a flexible and strong material, although it's not as hard as bone. However, there are some gene mutations that alter the appearance of the cartilage. There are two affected breeds, the aptly named American Curl and Scottish Fold, which, as their name suggests, have ear flaps with curls or folds. Fortunately, it doesn't seem to affect their hearing.

DID YOU KNOW?

All kittens are born deaf. Their ears are closed and they have no sense of hearing. Their ear canals start to open at five days, and by 14 days the kitten can work out the direction from which a sound has come. It takes four weeks for their hearing to develop fully.

INTENSITY OF SOUND

Have you tried sneaking up on your cat? It's really difficult. Not only do they have a better hearing range and swiveling ears that help pinpoint sounds, they are far more aware of the intensity of sounds than us. The loudness, or intensity, of a sound (the amount of vibration) is measured in decibels (dB). Humans can hear quiet sounds of around 10 dB, whereas, at the other end of the scale, 130 dB is a painfully loud noise. Cats can hear quiet sounds, too, but from a greater distance. In fact, a cat can hear a 30 dB whisper from a distance six times further than we can, so their hearing is perfect for detecting the squeaks and rustlings of a mouse. You will certainly never be able to get away with opening a bag of treats, even if you think the cat is upstairs asleep on the bed. However, at the other end of the scale, cats hear loud bangs with six times the intensity that we do, so it's hardly surprising that they are terrified by loud noises, such as fireworks and thunder claps. Given their sensitivity to loud sounds, you need to think about your cat when you play loud music, slam a door, or yell. The sounds will seem much louder to a cat and they can be easily scared.

DID YOU KNOW?

There is a high likelihood that a white cat with blue eyes will be born permanently deaf. The genes that cause the blue eyes and white hair are also linked to malformations in the cochlea, leading to deafness. Some white cats have only one blue eye and, if they are deaf, they are deaf on the side of the blue eye.

SPY-CATCHERS

Staff in the Dutch embassy in Moscow, Russia, noticed that the two embassy Siamese cats kept meowing and clawing at the walls of the building. The staff thought they had a mouse problem, but on investigation they discovered microphones hidden by Russian spies. The cats could detect the high-pitched squeak produced by the microphones when they turned on, a sound that the staff could not hear.

CHAPTER 3: THE NOSE KNOWS

Imagine you are a cat walking down a busy street. Your sight's not that great. In fact, everything close up is a bit blurry, but you do have a super sense of smell. You can detect smells that you never knew existed. You navigate by smell, finding your way home by remembering where you smelled freshly made coffee, the acrid aroma of gasoline, and the stench from the trash cans. Welcome to the cat's world of scent.

COMMUNICATION SKILLS

As well as using smell to find their prey and to check that their food is safe to eat and water clean enough to drink, cats rely on smell to learn about their environment. It's a two-way exchange: they leave scents for others to find, and they learn from the scents left by others. As they walk around their territory—the backyard, the urban street, or the rural field—they rub against posts to leave behind a calling card that tells other cats that they are there. Much of a cat's social behavior is influenced by smell, and it plays a key role in stress and anxiety reduction and helps cats to feel more secure in their homes.

SUPER NOSES

You have been using a brand of cat litter for years and, all of a sudden, your cat decides she's not going to use the litter box. The litter looks and smells the same to you, but your cat can detect a tiny change in the ingredients and she doesn't like it! This is just one way that our super-sensitive cats react to smells.

We have around five million sensory receptors in our nose that detect smell. Research by Leslie Vosshall at the Rockefeller University in New York has shown that humans have a much better ability to distinguish between scents than previously thought. So imagine what a cat can do with 80 million receptors—16 times as many. This gives them the amazing ability to distinguish between hundreds of thousands of different smells. And to boost their sense of smell further, they have a special olfactory organ called the Jacobson's organ (or vomeronasal organ) in the roof of their mouth. Many mammals have this "extra nose," but we don't. No wonder their world is dominated by smell.

HOW DOES THE NOSE WORK?

The cat breathes through her nose, drawing air into the lungs. As air passes through the nose, scents in the air mingle with mucus that coats the surface and protects the delicate nerve endings of the sensory receptors. The presence of scent molecules triggers a response and a message is sent to the brain.

The cat has a small, flattened, hairless nose supported by a bony case and some flexible cartilage. Deep in the nose's interior are many fine, scrolled bony plates called turbinates, which are covered by a thin layer of mucus. It's here that smells are trapped. Scents are carried through the two nostrils to the nasal cavity, which is divided into right and left sides by the nasal septum. There are hundreds of different kinds of sensory receptors, each able to detect a particular smell. When they are stimulated, a message passes along the olfactory nerve to the olfactory center in the brain, where it is processed and a recognizable "picture" of the smell is built. The Jacobson's organ lies in the roof of the mouth and is connected to the nasal cavity by two narrow tubes.

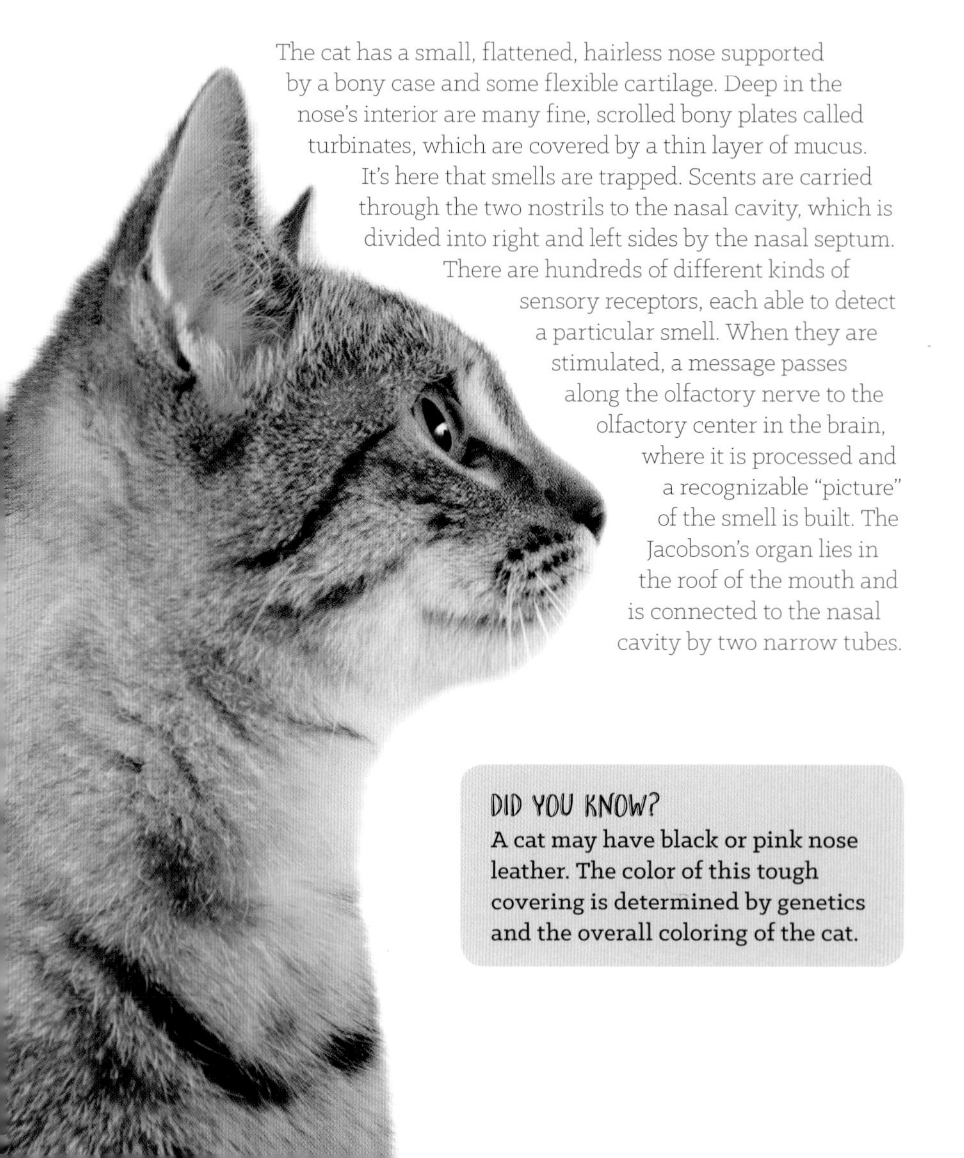

DID YOU KNOW?

A cat may have black or pink nose leather. The color of this tough covering is determined by genetics and the overall coloring of the cat.

SUPERIOR SMELL

For someone who isn't a cat lover, a cat's habit of rubbing himself on the furniture or, even worse, up and down a person's legs, is really annoying. But to a cat it's essential. Cats have scent glands around their face, and rubbing furniture and bedding leaves their scent everywhere and makes them feel secure. Changes make them anxious. For some cats, the worst kind of owner is somebody who keeps the house spotless, continually removing their reassuring scent and replacing it with a strange one.

FLEHMEN RESPONSE

Does your cat make a peculiar face, with her mouth slightly open, upper lip pulled back, and nose twitching? She's not grimacing at a bad smell but tasting the air. Called the Flehmen response, this behavior is seen in all cats, including the tiger. To analyze a smell, the cat takes a deep breath of air and squishes her scent-rich saliva all over the Jacobson's organ in the roof of her mouth. Cats are thought to use this Flehmen response to detect pheromones, and it may be especially useful for tomcats seeking females in season.

RECOGNITION

We recognize people from their visual appearance, relying on our sight to identify facial features and other parts of their body. Cats are different; they rely mostly on smell to recognize other cats and people, and a change in the smell of another cat can really throw them. Sometimes, it even leads to fights. If you keep more than one cat, I'm sure you have noticed that there can be some friction between them when you bring one cat home from the vet. There's definitely a bit of tension as the two are reintroduced. It's all to do with the returning cat having picked up lots of new smells at the vet, so the other cat thinks he is a stranger. There's a simple solution. Take a blanket or towel from home with you when you go to pick up the cat from the vet, and wipe the cat down with it before you return home, so the cat has more of the "home smell."

The same applies when you are introducing a new cat to a household that already has a cat. A newcomer is usually kept separate for a few days to get used to the smells of the other cat, but you can help by rubbing a blanket or toy from one cat over the other, and vice versa, so they get to recognize each other's scent.

Surprisingly, not all cats are highly smell-focused. Research by the University of Lincoln, in the UK, shows that, while most cats react to smells more than visual cues, a small number are more visually orientated. This accounts for the differences in behavior between cats in the same household, with some being upset by a change in the house smell, such as the use of a new floor cleaner, whereas those relying more on visual cues are not so bothered.

SNIFFER CAT

Rusik was a stray that was adopted by the customs guards at the port of Stavropol in Russia. The cat was found to have a talent for sniffing out caviar, a much-prized and very expensive luxury food made from the eggs of the sturgeon, a type of fish that is endangered. Valuable caviar from poached fish was smuggled through the port and the guards used sniffer dogs to find any caviar concealed in vehicles. Rusik soon replaced the sniffer dogs, as his ability to locate stashed contraband was much better than the dogs. Sadly, he was run over and killed, and many claim that it was a "hit" by organized crime!

DID YOU KNOW?

Smell is the very first sense used by newborn kittens. Their eyes and ears are closed, so they rely on smell to lead them to the milk bar. They soon learn to recognize and avoid offensive smells, too.

LEAVING A SMELL

The yowling of the local tomcat outside the window at night is a sure sign that your female cat is in heat. But how on earth does he know it is the right time to be serenading his potential mate? It's all to do with pheromones. These are chemical signals produced by cats to alter the behavior of other cats. Cats produce a number of different pheromones, but the best known are the sexual pheromones. A female cat produces a sexual pheromone when she is ready to be mated. She doesn't have to produce much; it is carried on the wind, and a tomcat can detect it from several miles away.

The so-called "facial pheromones" are left on household surfaces when a cat rubs his face along them. Over time, the regular rubbing leaves an oily brown mark. Other glands on the back, tail, and paws leave scents, while a nursing female also produces pheromones from glands around her nipples. This acts to calm down her kittens and to help her to identify them, should they become separated.

If you have owned an anxious cat, it's quite likely that your vet will have recommended the use of synthetic pheromones to calm him down. These pheromones mimic the scents of the facial pheromones, which are linked to a feeling of contentment. They may also help to stop your cat from chewing, scratching, and spraying.

DID YOU KNOW?

Mice and rats don't like the smell of cat urine and they stay away from it. But if a rodent gets infected with the parasite that causes toxoplasmosis, its behavior changes and it is attracted to urine. This increases the chances that it will be preyed upon by the cat and that the parasite will be passed on to the cat. The parasite actually changes the rodent's behavior in order to complete its life cycle and secure its future!

PHEW—TOMCATS!

The pungent urine produced by the tomcat is enough to drive some cat owners mad and it is the main reason why so many tomcats are neutered. But why does their urine have such a strong smell? It's all to do with communication. All cats add a bit of pheromone to their urine, but the urine of the tomcat is rich in a particularly strong-smelling amino acid called felinine. He squirts two or three powerful jets of urine over walls, fences, and gates as a sort of "calling card." The urine is oily, too, so it sticks, and, just for good measure, it also contains some secretions from the anal glands. There is no way another cat can miss this! As the tomcat does his rounds, he smells the various calling cards to work out which other cats have been in the area, what sex they are, and how long ago they passed by.

HALLUCINATING CATS

If you didn't know what was going on, you would think that the sight of your cat rolling, rubbing, licking, and even salivating over a nondescript gray plant in the backyard meant he had lost his mind. And you wouldn't be too wrong. The smell of the aromatic catnip (the scientific name is *Nepeta cataria*) can drive some cats completely mad with pleasure, so much so that they can lose themselves for a while and even become completely unaware of everything going on around them. And then it suddenly wears off. They get up and walk away, just as if nothing had happened. The reason? Catnip is a bit like marijuana to a cat and they are highly stimulated by it.

Catnip is a member of the mint family and it's rich in volatile oils and tannins. There is one oil in particular, called nepetalactone, that enters the cat's nose and stimulates some of the sensory receptors, provoking a strong sexual response in the brain. In fact, this volatile oil mimics the cat's sex pheromone. Not surprisingly, catnip is used in many cat toys. And it's all to do with genetics. If a cat has the gene, he will respond to catnip; if not, there will be no reaction. The gene is found across the cat family, so a lion is just as likely to go mad over the smell of catnip as your pet cat.

CHAPTER 4: TASTE

Cats are often very fussy eaters, liking one food but not another. Or one day they like a food and the next they don't. We try all sorts of ways to get them to eat, but the reason behind this behavior may lie on their tongue.

TASTE BUDS

A cat's tongue is covered with tiny projections called papillae (see the photograph at right), which create a rasping, rough surface that is perfect for scraping meat off bones and for grooming. Taste buds are found in the papillae. A human's tongue is covered with more than 9,000 taste buds, whereas a cat, thanks to her much smaller tongue, has 500. The actual receptors on the taste buds are formed from protein that lies on the surface of the tongue. When the protein comes into contact with specific dissolved chemicals in the food, the receptor is activated and a message is sent to the brain.

Humans have at least five types of taste bud on their tongue—salty, sour, bitter, sweet, and umami, and there may even be a sixth that responds to fatty foods. Cats are different. They've got taste buds, but they have fewer types. We love sweet-tasting foods and so do lots of other mammals, but cats are unusual in that they lack the ability to detect sweetness. As far as we can tell, they can detect only protein, bitter, and fatty tastes.

DID YOU KNOW?

Another oddity in the cat's sense of taste is his preference for warm food. While we love ice creams and sorbets, the cat has no taste for chilled foods, preferring food to be at body temperature—ideally about tongue temperature. This probably relates to eating the warm bodies of prey. If you are having problems with your cat eating his food, don't give him food straight from the fridge—try warming it up a little.

SWEET TASTES

Why can't a cat taste sugary foods? The answer lies in genetics. The ability to taste sweetness comes from two genes that are responsible for the manufacture of two specific proteins, which lock together to form a taste receptor.

Most animals find it useful to know that food contains sugar. The presence of a sweet taste tells the animal that the food is energy rich and that it contains readily available sugars. It's a great energy source that can be used straight away. Humans eat a mixed diet, one with both animal and plant-based foods, as do many other animals, such as bears, wolves, and dogs. All have the two genes and can taste sugar. Even grazing mammals, such as antelope and zebra, can taste sugar.

But the cat is different. Cats, both domesticated and wild, are meat eaters, but unlike the wolf they do not eat berries and other plants. They are known as "obligate carnivores," living solely on meat. So, it is not surprising to learn that one of the genes controlling the sweet receptor protein is inhibited. Scientists working at Monell Chemical Senses Center in Philadelphia found that the gene has been altered in the past, so it cannot function and the cat is unable to make this protein and thus lacks a sweetness receptor. Further studies have shown that cats aren't the only obligate carnivores that lack a craving for sugar. They have been joined by spotted hyenas, sea lions, seals, and dolphins.

Studies have shown that, as well as not being able to taste sugar, cats lack a key enzyme in the liver that regulates the amount of sugar in the body. So it seems odd, therefore, that many commercial cat foods contain as much as one-fifth carbohydrate, which the animals can't deal with properly. This may well account for the rise in the number of cats suffering from diabetes.

ENERGY FOODS

Cats may not be able to taste sugar, but they can taste a substance called ATP—adenosine triphosphate. ATP is an energy-rich molecule found in every cell; it's a type of energy currency. Once inside the cell, sugar is broken down in a series of carefully controlled steps. This releases energy that is then locked up in ATP. If a food, such as an animal's muscle, is rich in ATP, it means it is probably a good source of energy.

A BITTER TASTE

Bitter is one of the tastes that both cats and humans can recognize, but cats have a far superior ability to detect the taste than us. They are "bitter super-tasters." But why should they be so much more sensitive to a bitter taste? Among the mammals, herbivores are able to taste bitter substances, and this is thought to enable them to detect the bitter taste of many toxic plants. But the reasoning behind an obligate carnivore having bitter receptors is less clear.

Studies by biologist Dr. Peihua Jiang of the Monell Chemical Senses Center found that cats could use their bitter receptors to identify potentially poisonous prey, such as frogs and toads, or even detect plant toxins in the gut of their plant-eating prey. Another line of investigation is looking at the role of bitter receptors in the detection of infection. These receptors are found not just on the tongue but also in organs such as the lungs. Scientists think the cat has more bitter receptors than she would need solely for tasting as part of her defense against bacterial infections.

The presence of bitter detectors may help to explain why cats are such fussy eaters. With their sensitivity to bitter tastes, they are more likely to detect unpleasant tastes in any food that we give them. With owners feeding their cats highly processed kibble that looks nothing like their natural food, it's perhaps unsurprising that cats recognize tastes that we are completely unaware of and find it very off-putting.

CHAPTER 5: A SENSE OF TOUCH AND BALANCE

Ask a cat owner what the whiskers do and they'll probably say, "They help a cat figure out whether or not it can get through a gap." And, yes, they do, but they also do lots of other things as well.

THE CAT'S WHISKERS

Whiskers, or vibrissae, are important touch receptors. Cats have 24 movable whiskers, up to 12 on each side of the face, arranged in three rows. Each whisker is long and thick, with deep roots, and is connected to a muscle so that it can be moved backward and forward. There is a cluster of nerve endings at the base of the whisker and there are sensory receptors at the free end of the whisker. These receptors are sensitive to changes in the surrounding air, such as tiny air currents and changes in air pressure and temperature. Any movement of air around the face pushes on the whiskers and even a minute movement is enough to set off the receptors and send messages to the brain. Since the end of a whisker is so sensitive, the cat knows instantly if it touches an object, so the whiskers help the cat to work out the width of a gap and to find his way about and ensure that, even in the dark, he does not bump into anything.

Whiskers, however, are not the only sensory hairs on a cat's body. His fur is generally sensitive to the touch, and there are short, stiff sensory tufts on his eyebrows, cheeks, chin, and elbows. There are also short whiskers on the back of his front legs that help him to put his feet down in the right place without having to look down. There are also a few sensory tufts in his ear flaps. Finally, his pads are incredibly sensitive, as are his paws, and even his canine teeth have a touch-sensitive surface.

DID YOU KNOW?

It is cruel and dangerous to trim the whiskers of a cat, because you are removing their ability to touch. The loss of these sensory whiskers leaves a cat disorientated and often frightened. We don't realize just how important a cat's whiskers are in the dark and how they help her to "see" in front of her. If a cat is blinded, either through disease or injury, she will use her whiskers as substitute eyes.

WHISKER MESSAGES

The information provided by the whiskers is often combined with that coming from other senses. For example, information from the whiskers is sent to the inner ear, where the information is used by the balance center. Due to poor close-up vision, a cat uses his whiskers to work out what's going on in front of his nose. When he grabs small prey, such as a mouse, the whiskers are swept forward to form a basket around the animal and information is sent to the brain about its size and shape. Critical information concerns the position of the nape—the point on an animal's neck that is most vulnerable to a killing bite. Cats learn to kill when young, and their most important move is a bite with their powerful jaws at the nape of the prey's neck, whereby their canines slide between the vertebrae of the spine and sever the spinal cord.

Whiskers are important in communication, too. The position of the whiskers, either lying flat or twisted forward, sends clear messages to other cats and even to people. A cat with his ears flat and his whiskers lying against the face is sending a message to stay away, while whiskers twisted forward indicate that the cat is interested and friendly.

DID YOU KNOW?

The pattern of whiskers along the side of the face is unique to each cat and represents something of a feline "fingerprint."

BALANCE

How many times have you heard it said that cats always land on their feet? Thanks to their amazing sense of balance, which enables them to walk along fences and branches with ease, few of us have seen a cat crash-land. But when they do fall, their balance means that they land on their feet.

Balance is the second function of the ear. The cochlea is not the only part of the inner ear. Alongside the cochlea are three fluid-filled tubes called the semicircular canals. They are set at right angles to each other and, when the cat tilts her head to one side, fluid inside moves around, pushing on sensory receptors that send messages to the brain. Each of the three semi-circular canals picks up movement in a different direction, so the horizontal canal is sensitive to the head being spun, while the others pick up on side-to-side movements and up-and-down movements. There is always a bit of inertia, as the fluid doesn't stop moving immediately, which accounts for dizziness.

A cat's long tail is an important balancing structure, too. It helps the cat to balance as he walks along a fence or wall, swinging from side to side as a counterweight. If the cat starts tipping, his tail moves in the opposite direction to his head, so keeping the body balanced.

LANDING ON THEIR FEET

It's amazing how cats can squeeze through the narrowest of gaps. This is due to their flexible back (which has five more vertebrae than our own) and the lack of a proper collarbone. Not only does this mean that they can wiggle through tiny gaps, but their body can also rotate more easily in the air. This helps them to twist as they fall, so they are more likely to land the right way up. This maneuver is called the "righting reflex."

When a cat falls off a wall, for example, his back is likely to be pointing to the ground, so he will bend his body and twist his head around. This is followed by his front legs twisting and then his back legs, which results in the cat righting himself in mid-air. He will ensure that his front legs hit the ground first, followed by the back legs, and he'll arch his body, too, as this helps to absorb the shock of landing. The tail is also involved, moving around and keeping the cat level in the air. The cat will raise his tail vertically to stop forward movement and assist with the landing.

DID YOU KNOW?

Cats are born with the ability to right themselves; it's inbuilt and doesn't have to be learned. By the time a kitten is six weeks old, the righting reflex is fully developed.

SURVIVING A FALL

There are many accounts of cats falling out of windows in high-rise apartment buildings and surviving the fall to the ground. One particular cat, unsurprisingly nicknamed Lucky after the event, fell from a window 26 stories up a skyscraper in Manhattan and, apart from a few broken bones, survived the fall. Another remarkable cat, called Sugar, fell from a 19th-story window in Boston and suffered only a few bruises.

Cats have long lived in trees, so being able to survive a fall to the ground is a vital survival adaptation. As well as the ability to right themselves in mid-air, cats have other ways of slowing down the rate at which they fall from great heights. They have lightweight bodies and thick fur and, as they fall, they spread out their legs to increase drag, rather like a parachute, which reduces their rate of descent. They relax before impact, which helps to lessen the severity of the landing, while their long legs act as shock absorbers. Studies have found that the terminal velocity of a cat falling to the ground is 60 mph (97 km/h), whereas an average-sized human would fall at speeds of up to 120 mph (193 km/h), so a cat is more likely to survive a fall than a human.

HIGH-RISE SYNDROME

Working in Manhattan, vets Wayne Whitney and Cheryl Mehlhaff were used to seeing cats being brought into the surgery after falling out of windows, victims of the so-called "high-rise syndrome." In 1987, they decided to conduct a study into the injuries sustained by cats after a fall. Over a five-month period, they studied 115 survivors that had fallen between two and 32 stories onto the pavement below. The average height was five and a half stories. Of these cats, 90 percent survived. Unbelievably, one cat that had fallen 32 stories not only survived but two days later was fit enough to go home. The vets' statistics revealed some extraordinary observations. Cats that fell between seven and 32 stories were actually less likely to die than those falling from two to six stories. The reason that the cats that fell farther survived may be to do with them having more time to right themselves and slow down their descent. Of course, we have to bear in mind that any cats that died as a result of their fall didn't make it to the vet center, while cats falling out of the lower stories may not have been badly injured and did not get taken to the vets for treatment.

SECTION TWO
CAT INTELLIGENCE

We know that cats are an intelligent bunch, but they use their intelligence in a different way to us. Living mostly outdoor lives, they have learned to fend for themselves; they are loners, prowling the streets and fields on their own, confidently finding their way around in the day or at night. They can identify hundreds of different smells and understand what makes other animals tick. They know, for example, that rats and mice spray urine over their feet to leave a smelly trail, and they use this knowledge to hunt them.

The modern city cat must be particularly street-smart, and she must use her intelligence to avoid multiple dangers, such as cars, other cats living nearby, and even dogs. Highly intelligent and curious in nature, cats can learn tricks, such as opening doors, and they can even be taught to flush the toilet.

CHAPTER 6: THE PROCESSING CENTER

Key to the cat's intelligence is his brain. Measuring about 2 inches (5cm) long and weighing around 1 ounce (25g), a cat's brain occupies proportionally less space relative to his body mass than that of a human—about 0.9 percent of body mass, whereas the human brain accounts for 2 percent of a human's total body mass. But, in the area where information processing takes place, the cerebral cortex, cats have lots of nerve cells. In fact, cats have almost twice the number of a dog, a whopping 300 million neurones, compared with the dog's 160 million. It is not surprising, then, that cat owners think their pet is much smarter than a dog.

However, cats' smaller brains mean that they have limited space for storing memories or problem solving. They are more likely to live for the moment and react to incoming information in a simple way. But their visualization of a 3D environment, such as their backyard or street, is fabulous, and this allows them to form mental maps of their territory to work out the best routes around it and where to find prey.

DID YOU KNOW?

The brains of cats and humans have gray and white matter. The folded surface of the cerebrum has a gray-brown appearance. Underneath is the white matter.

The brain of the cat and human is similar, but there are differences that relate to the processing of sensory information and the long-term storage of memories.

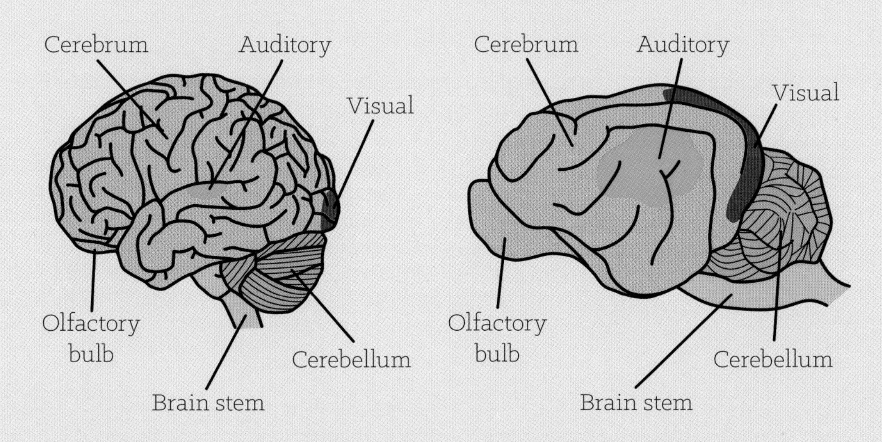

Cerebrum: This is the largest part of the cat's brain. The outer layer is called the cerebral cortex, which is divided into a right and left cerebral hemisphere. There are large areas devoted to visual and auditory information and there is a distinctive olfactory bulb.

Olfactory bulb: This is the center for smell, and it is particularly large in the cat.

Cerebellum: This is found under the cerebrum. It is involved with coordination of movement, posture, and balance.

Brain stem: This connects the cerebrum and cerebellum to the spinal cord. Its role is in the automatic functions of the body, such as breathing, heart rate, body temperature, and digestion.

THE CEREBRAL CORTEX

When you look at the brains of cats and humans, you can see right away that the cat's cerebral cortex is not as highly folded as that of a human. Both have two cerebral hemispheres, each of which has four lobes: the frontal, occipital, parietal, and temporal lobes. Each lobe has a specific set of functions.

Frontal lobe: This deals with behavior and emotions, intelligence, problem solving, vocalization, and movement of the muscles.

Occipital lobe: This is responsible for the interpretation of visual signals.

Parietal lobe: This is the visual center, and deals with the interpretation of information from the ears and skin, as well as with temperature control, motor area, and memory.

Temporal lobe: This relates to hearing, memory, and organization.

The cat's brain works in much the same way as our own. There are different areas performing specific tasks and all the areas are well connected, so there is plenty of information sharing. This enables the cat to take all the information from his senses, process it in super-quick time, and react to the environment. For example, the cerebral cortex receives sensory information from the five senses, interprets the information, and stores some of it in the memory areas. It also sends information to the motor area, which controls the cat's body movements.

VISUAL CENTER

Cats have more nerve cells in the visual processing areas of the cerebral cortex than humans and most other mammals. This is where all the information carried by the optic nerves ends up, where it is processed and compared with visual memories.

LIGHTNING REFLEXES

The amazing ability of cats to twist as they fall, and so land on their feet, is due to a reflex. Reflexes are automatic processes that take place instantly and without any conscious input. A reflex involves the stimulation of one or more senses, which sets off a series of reactions. First a message speeds down a nerve to the brain or spinal cord, which in turn kicks off an instant and automatic response in another nerve, which causes a muscle or set of muscles to move. Blinking, coughing, and recoiling from a hot object are all reflexes. Another example is the flight-or-fight response, an automatic protective behavior that kicks in without us having to think about it. It's present in the cat, too, an immediate response to save her life, such as with the righting reflex or the pupil closing instantly to prevent too much light entering the eye.

DO CATS HAVE FASTER REFLEXES THAN US?

They probably do, but it's not due to their nerves sending messages faster. It's simply that, being smaller animals, the messages have less distance to travel, so a message takes less time to travel from the sense organ to the brain.

Brain-scanning technology is improving and, hopefully, it will enable researchers to work out which areas of the brain are activated in response to certain cues, such as particular scents or pheromones. We know already that a region deep in the cerebrum, called the caudate nucleus, is linked with the processing of smells. Recently, researchers trained dogs to lie still in an MRI machine so that they could scan the dogs' brains and see which areas respond when the dogs are given certain stimuli. In one test, the dogs were presented with different smells, and the MRI scan showed that their olfactory bulbs were activated by all smells, but the caudate region was activated by the scent of a familiar person, such as their owner. It has not been possible to carry out similar types of experiments with cats, but researchers don't think that cats will differ that much from dogs, so they are likely to be able to distinguish between people on scent alone.

The caudate nucleus has an important role in learning, especially the storing and processing of memories. When new information comes in, a cat uses information stored from previous experiences to make decisions about the appropriate responses to make. So information about the smell of the owner would be in the memories of the cat and would trigger a response when the olfactory information reaches the caudate nucleus.

DID YOU KNOW?

The computer giant IBM simulated the brain of a cat, a feat that required 25,000 processors. To simulate a human brain, they would need at least 880,000 processors.

EARLY DEVELOPMENT

The first few weeks in the life of a kitten have a lasting effect on the way her brain develops. Among the first researchers to explore brain development in kittens were Colin Blakemore and Graham Cooper in 1970. In a classic experiment, two groups of young kittens were placed in an environment that was covered with either vertical or horizontal lines, restricting their visual experiences for a number of hours per day. It was shown that the group that was raised with vertical lines was unable to detect horizontal lines, and vice versa.

This simple experiment showed the critical role that the environment plays during development. It was clear that the brain is not hardwired at birth, but is instead relatively plastic. The visual center of the cerebral cortex was able to undergo change during a critical period after birth and adapt to the environment.

In other experiments, researchers found that cats raised in a laboratory-type environment with few visual stimuli had poorer visual perception than cats that were able to roam outside. Once again, the research showed how important it is for kittens to be given a rich environment for those first few weeks of their lives.

LOOKING FOR MILK

Where is the milk? The first life challenge faced by a newborn kitten is to find her mother's nipple and take her first drink of milk. The mother cat licks her newborn kittens clean, but her kittens are on their own when it comes to finding milk. They have to use their initiative to make their way to the nipple, and at the same time to compete with the other kittens to get there first.

This is a time when instincts take over. The newborn kittens are blind and deaf, so they use their sense of smell and touch to navigate to the row of nipples on their mother's underside. Once they are in the right area, kittens go into a "nipple-search" mode, moving their heads back and forth to locate a nipple and latch onto it. Kittens tend to use a particular nipple, with the stronger kittens acquiring rights to the best nipples. It's likely that each kitten asserts her right to a nipple by smearing her odorous saliva over it as a kind of chemical signature.

Smell has other uses, too, in these first vital weeks. Kittens have to be able to find their way back home should they stray, and for the first 12 days or so, their eyes are closed, so their sense of smell takes center stage. The mother makes sure that the nest area is smeared with her smells.

DID YOU KNOW?

In cats that are born permanently blind or suffer a brain injury, other parts of their brain assume the function of the damaged part.

MEMORIES

We know cats have memories, just as humans do, and that their memories help them to make decisions and learn. Memory is complex and involves not only short- and long-term memory, but also a third type, known as skill memory. Short-term, or working, memory occurs in the frontal cortex. Information is stored for up to one minute and there is a limit to the number of items stored; in humans it is around seven. For example, we use our short-term memory to memorize a telephone number, whereas a cat may use it to memorize the position of a treat.

Long-term memories are stored in the hippocampus of the temporal lobe. It is here that the memories are stored for a much longer period; some for life, and others for only short periods. While our long-term memories have unlimited content, we are not sure about those of the cat and whether or not there is a limit to the number of memories, since their brain size is smaller. However, we do know that their long-term memories can last for years. For example, your cat may well remember where food or a toy was hidden some years earlier.

The third type of memory, skill memory, is located in the cerebellum, and it is here that we find the stored automatic learned memories. For humans, this includes memories such as "how to ride a bike," a skill we learn and then never forget. For cats, their skill memories are vital for their survival.

Long-term memory is key to problem solving. We learn and store information so that we know how to deal with the same thing in future. Cats, with their much smaller brains, have a limited amount of space available for memory storage. They can react to incoming sensory information and have some basic memories, and they can undertake simple mental processing. But it is therefore much more likely that cats live more in the present and the recent past.

DID YOU KNOW?

It is perhaps unsurprising to learn that diet can affect not only a cat's physical health but also his cognitive condition, including his memory and capacity for learning.

DOES YOUR CAT HAVE A GOOD MEMORY?

There are differences between individual cats. Just as with humans, some cats have a better memory and good learning ability, meaning that they can remember where food has been hidden and learn new tricks, whereas others are less able in these areas. Breeds, too, have significant differences. Abyssinian and Siamese cats are known to have a good memory, while the Scottish Fold is adaptive and can be taught tricks, such as fetching toys and walking on a lead.

OBJECT PERMANENCE

This is a fancy term for the ability to keep something in mind when it disappears. For example, when a ball rolls under a piece of furniture, we can't see it, but we still know it's there. This involves working short-term memory and we see this cognitive ability appear in toddlers of just two years of age.

But do cats show this ability? The answer is yes. If a kitten sees his favorite toy disappearing under the sofa, he stays there looking, despite no longer being able to see it. This ability is essential for hunting, because prey animals will disappear behind trees or into recesses in the ground. The cat will remember where the animal disappeared and search for it there and, if that proves unsuccessful, will enlarge the search area. This is when a mental map of their territory comes into play, as they know all the possible hiding places.

Researchers can test this ability quite easily. They place some cat food in a bowl positioned so that the cat can see it and then place a screen between the cat and the bowl, obscuring the cat's view of the food. The cat's ability to keep the object in mind means that he is quite likely to go behind the screen to find it. The trouble is that cats are easily distracted and they may find the object that is hiding the food more interesting! There are some tests you can try with your own cat in Chapter 10.

How long does a cat hold on to these working memories? In a simple test called a delayed response task, cats are encouraged to hunt for a toy that has disappeared. Researchers initially demonstrated that cats have a working memory of about 30 seconds, which shortened when the cats were given more choices. A subsequent test revised this upward. A cat was held by their owner and allowed to see an object, which was then removed. After a certain number of seconds had elapsed, the cat was released and allowed to run over and look for the object. The researchers found that the working memory lasted for as long as 60 seconds.

THE AGING CAT AND DEMENTIA

Older cats suffer from cognitive dysfunction, which is a bit like dementia in people. Like humans, cats lose brain cells as they age and this is likely to affect their memory. They may have a reduced ability to learn new skills and may exhibit poorer memory. Memory in cats can also be damaged by disease. People suffer from Alzheimer's disease, whereas cats suffer from Feline Cognitive Dysfunction. The symptoms include disorientation, increased vocalization, avoidance of social interactions with people and other cats, poor sleep patterns, house soiling, and irritability. Sadly, there is no cure, but vets can ease the symptoms with medication.

INNATE BEHAVIOR

Cats are natural-born hunters, but what does this really mean? Cats don't have to learn to hunt, because this is an innate behavior that they are born with. In other words, it's a natural or instinctive behavior. Despite thousands of years of domestication, the instincts of a wildcat persist and lurk just below the surface in pet cats. They have retained a strong hunting instinct and, much as we try to stop our cats from catching birds, we usually fail.

Animals don't have to practice a natural behavior repeatedly to get it right. They instinctively know what to do. This type of behavior is predictable, too. The sucking reflex seen in kittens and human babies, where they will instinctively suck on a nipple or a similar-shaped object placed in their mouth, is another innate behavior. The reason for it is simple: to increase their chances of survival.

LEARNED BEHAVIOR

As a cat gets older, he learns or acquires new behaviors. Once again, this starts as a kitten. There is a lot of socialization between the kittens in a litter. They learn through watching and playing, finding out how roughly they can play with their siblings, and how hard they can bite or knock one another over. They learn from their mother, too. She may reprimand them when they bother her for milk after she has weaned them, so they learn which behaviors are acceptable and which are not.

Additional behaviors are learned as a cat gets older, and he may begin to associate his behavior with consequences, such as meowing at his owner to prompt them into giving them a treat. He may come to associate the sound of his owner using a can opener to open his canned food or the rattle of kibble being tipped into his bowl as a sure sign that it is meal time.

Learned behavior must be practiced and perfected. We learn to ride a bike, while a cat learns to play with a food puzzle. What is important about learned behavior, when compared with innate behavior, is that is it adaptable and can be changed, which is important for survival. The more intelligent an animal, the better it is at learning and the greater its repertoire of behaviors. Innate behavior cannot be changed; the animal is locked into the behavior pattern. But the flexible and changeable nature of learned behavior is fundamental to their survival and the evolution of their species. For example, migrating animals are born with the desire to travel to summer or winter feeding grounds. They know what time of year they must leave and the route they must travel. This being innate behavior, it cannot be changed. But it does mean that, if something disastrous happens to the route, they can do little about it. For example, it could be as simple as migrating toads finding that a road has been built across their route to the breeding ponds. Their instincts take them across the road, even if they end up being squashed by cars.

Behaviorists categorize different types of learned behavior.

Habituation: This is a simple form of learning a new behavior. It's a behavior acquired after being exposed to a stimulus repeatedly. A new experience can change the animal's behavior. For example, if a cat comes across a new object in the home, her instinct, quite wisely, is to be wary. Some cats may even find the new object scary and run away. But, if they keep on coming across the object and nothing harmful happens, they decide that there is no danger and learn to ignore it.

Observational learning: Kittens watch and learn, especially when they are with their mother. They learn all sorts of behavior from her and also how to perfect other behavior. They may have an instinctive ability to hunt, but, by watching how their mother stalks her prey, pounces, and kills, they can copy and perfect their hunting technique. They watch their siblings or companions play with toys and they pick up new behaviors by copying them.

Conditioned behavior: This is a behavior that is acquired through reward. It is the way we teach our cats new behaviors, encouraging them to do something by giving them a reward when they do it successfully. There is an opposite way, too, through punishment. Cats are discouraged from performing an unacceptable behavior through being told off. You can read more about learned behavior in the next chapter.

DO CATS DREAM?

Have you seen your sleeping cat swish her tail, move her legs as if running, wiggle her whiskers, or chatter? Well, this is your cat dreaming. Cats spend a lot of time asleep—as much as 20 hours per day. Just like human sleep, cat sleep can be divided into two phases: REM (rapid-eye-movement) sleep and non-REM, or deep, sleep. We dream during REM sleep and cats do the same. The part of the brain that controls memories is the hippocampus and it's also involved in dreaming. Our dream phase comes around every 90 minutes, but cats have a dream phase every 25 minutes. What do they dream about? We can only guess, but it's probably about hunting.

Our brain has a kind of "off-switch," which makes sure that the muscles moving our arms and legs are not activated, so that we don't get up and act out our dreams. It's the same for cats and, during REM sleep, a cat is usually totally relaxed. But this off-switch may not work perfectly, and you may well see your cat's legs move and tail twitch. In older cats, this off-switch becomes less reliable and you may see them moving around in their sleep. Sometimes they even manage to wake themselves up with a start and look very confused and startled! The next time your cat falls asleep, watch out for signs of dreaming: REM starts about 15 minutes into sleep.

Given that we sometimes have nightmares, it's probably safe to assume that cats dream about their less pleasant experiences too. People recount how some rescue cats occasionally make horrible noises in their sleep and will wake suddenly with a fearful look on their faces. One video online shows a kitten having a bad dream, waving his paws around and, when this is noticed by the mother, she hugs the kitten and wraps her paw around his body.

DEEP SLEEP

Hardly anything happens in your home without your cat knowing. For about two-thirds of the time, your cat's senses remain active while he sleeps, and this means that he can respond to sounds and smells. No wonder you can never open your cat's food without him appearing, even if you thought he was asleep upstairs. But there are phases when your cat is completely switched off and takes a bit longer to come out of sleep. That's when you see him go through the process of waking up, yawning and stretching, and leg flexing, followed by a quick groom to freshen up before heading out once more into the world.

DID YOU KNOW?

Some cats may sleep for as long as 20 hours per day, although the average length of time spent sleeping is around 12 hours.

DREAM EXPERIMENTATION

We know quite a lot about dreams in cats from the work of Michel Jouvet, a French professor of experimental medicine. In 1959, he operated on a cat and removed the tiny bit of the brain that forms the off-switch for muscle movements during sleep. By today's standards, it was a barbaric experiment, but it revealed a lot. During the non-REM phases of sleep, the cats were quiet and slept peacefully with no body movement. When the REM phase started, they leapt up and started to hunt—stalking and pouncing on imaginary prey. At other times, they hissed and growled and arched their backs as if they were having an encounter with another cat. The research showed that, as suspected, cats do dream about things that happen in their lives, just as we do.

A SENSE OF TIME

Many owners claim that their cats have a good concept of time, knowing when their food is due, when to wake up, and so on. Many animals exhibit a similar ability, as they have a 24-hour internal clock that enables them to know when to become active and when to sleep. This is called the circadian rhythm and it is controlled by the so-called master clock, which is located in a part of the brain called the suprachiasmatic nucleus (SCN). The SCN is found in the cerebral hemisphere, close to the entry point of the optic nerve. Here there are thousands of nerve cells and they control patterns of sleep and wakefulness.

As well as having a good sense of time of day, it seems cats are pretty good at judging lengths of time. Being able to estimate short durations of time, say seconds or minutes, can be important in decision making and it has a role to play in survival. It is critical when chasing prey and judging the right time to make the all-important pounce. Researchers have found that cats are able to discriminate between five and 20 seconds. In fact, some cats can tell the difference between five and eight seconds. Time may play a particularly important role at night, when the cat is active in the dark, as it may help it to estimate distance and navigate home.

CHAPTER 7: HOW CATS THINK

We've kept cats around our homes for thousands of years, but, despite humans developing lots of different breeds of domestic cat, they're still pretty wild. People develop really close relationships with dogs, but, as cat expert Professor John Bradshaw explains, dogs recognize us as being another type of animal and behave differently with us than they do with other dogs. Cats, on the other hand, behave toward us in much the same way as they behave toward other cats. And we still don't know much about how they think.

FIRST LEARNING

A newborn kitten is particularly cute and vulnerable. He is born blind, deaf, and pretty much helpless, and he's entering a critical stage of his life, when he has to learn many of life's essential tasks. As soon as he can see, he watches carefully as his mother grooms herself, perfecting the technique on himself. He learns how to play with his littermates and the social niceties this involves.

This is the time that humans can really influence and mold a cat's personality. While it's important for kittens to learn how to behave around their mother and littermates for future interactions with cats, socialization with people is just as important. As the work of Roberta Collard and John Bradshaw has shown, a kitten that receives lots of handling will grow into a friendlier and less fearful cat.

THE IMPORTANCE OF HANDLING

Handling kittens can have a real influence on their future behavior. Research by Eileen Karsh and Dennis Turner in 1988 showed how a little bit of regular handling can have a huge influence on their psyche. In one experiment, kittens handled for 40 minutes per day were significantly more friendly and happier in their adult lives than those handled for just 15 minutes per day. In 2008, Rachel Casey and John Bradshaw showed that people buying a well-socialized cat got more emotional support from it and the cats displayed less fear of people. Interestingly, handling the kitten had other effects. Kittens handled for ten minutes twice per day matured earlier than those that had received no handling. For example, they opened their eyes sooner and were ready to leave their mothers earlier.

What does this mean to you or me? If you are buying a new kitten, it is really important to ask the breeder about the handling. Kittens raised outside in a shed or barn are less likely to have been socialized to the same extent as those raised in a family home, where they would have received lots of human interaction.

DID YOU KNOW?

The world's best mouser was a tortoiseshell cat called Towser (1963–1987). In her lifetime, she caught almost 28,899 mice. Towser worked for the Glenturret Distillery in Scotland and a statue has been erected in the distillery grounds to honor her.

A LOVE OF HUNTING

The pet cat doesn't want for much; there is always a supply of food (sometimes too much), water, a warm place to sleep, and lots of toys. What more could she want? The answer is to hunt! The desire to hunt runs very deep, and the sheer enjoyment that it evidently gives them means that kittens simply can't resist practicing stalking, chasing, pouncing, and killing. This behavior starts early, and techniques are continually refined as the cat grows up.

We see this in the big cats, too. In a pride of lions, for example, young cubs are brought food, but, once they are older, they are allowed to accompany the pride on a hunt. At first, they watch the adults hunt from a vantage point and then they are allowed to come to the kill to feed. Gradually they learn more of the hunting process and, eventually, they are allowed to join in.

Why do our cats love hunting? The excitement of hunting results in the release of feel-good hormones called endorphins. It's just like people getting pleasure from exercise and it can be an addiction. The cat continues to want to hunt and kill, and demands to be let outside at night. If this is restricted, they can become frustrated and need to be given other outlets. This is where toys play a role; furry, squeaky toys are preylike, and so help cats to vent their frustration.

SHOULD WE BREED OUT HUNTING?

John Bradshaw believes that we should. He thinks that we need to breed out the desire of the pet cat to hunt. The pet cat receives plenty of food and very few are still used for their role of pest control on farms. He believes that, as the world gets more crowded and wildlife is put under even more pressure, we cannot afford for the pet cat to kill millions of animals each year for, essentially, no good reason. Some of the more recent breeds, such as the Bengal (a cross between the domestic cat and an Asian leopard cat), have a very strong drive to hunt and can be aggressive to other cats. Siamese, on the other hand, have been bred over generations to be the perfect house cat and have all but forgotten how to hunt. Bradshaw would like to see people breeding for cats that don't hunt. Meanwhile, we have to deal with our natural hunters and one way can be to fit bells to the cat's collar so that birds and rodents are forewarned that there is a cat on the prowl.

WHAT DOES YOUR CAT GET UP TO AT NIGHT?

Do you know where your cat goes when he's outside at night? Does she stay in your backyard or does she wander far? In 2013, researchers from BBC's *Horizon* program teamed up with Professor Alan Wilson from the Royal Veterinary College, in the UK, to find out what cats got up to while they were outside. The team used safety-release collars fitted with GPS, which was active while the cat was moving. They also fitted a small radio transmitter, so they could recover the collar should it fall off. The team was able to record the movements of the cats in the study, all of which lived in a small village in Surrey, England. Then they processed the data from the GPS trackers and used software to overlay the movements on aerial photographs, so they could visualize the cats' routes. They found that some cats stayed close to home, while others roamed into the surrounding countryside. A few visited the homes of neighboring cats, while others would avoid meeting other cats.

Similar studies are taking place elsewhere in the world, the largest and best known of which is Cat Tracker. This study was set up by researchers based at North Carolina State University. More than 500 cat owners have signed up already to Cat Tracker, the majority living in the USA. As of early 2017, the project is still live, with recorded tracks from more than 100 cats. The University of South Australia joined the project in 2015, with a team led by Dr. Philip Roetman from the Discovery Circle. As well as helping us to understand how far our pets roam at night, Cat Tracker will help local authorities manage problems associated with cats, such as the hunting of birds and other wild animals, fighting, calling at night, and spraying and defecating in public parks.

Volunteers taking part in Cat Tracker need a GPS harness. If they live close to either of the universities involved in the project, they can loan the equipment. If not, they purchase a recommended model. The cat wears the harness for five days. After five days the GPS unit is removed but the harness is left in place. The data is uploaded to the researchers and then recharged and fitted on the cat again for a second five-day tracking session. Initial results from the US project showed that most urban cats traveled over an area of less than 12 acres (5 hectares) and generally they kept within their built-up neigborhood, rather than venturing farther afield. Just 5 percent of the cats traveled relatively far, with the most adventurous cat exploring an area of 116 acres (47 hectares).

More than 400 cats have been tracked in South Australia, where researchers found that cats ranged up to 76 acres (31 hectares), although the average area covered was just 2½ acres (1 hectare). The majority of cats wandered farther at night than during the day. Just 3 percent of the cats ranged more than 25 acres (10 hectares). Male cats had a bigger range than female cats. They found there were two types of cats: sedentary and wandering (cats that ranged over 2½ acres). These wanderers crossed more roads, had more fights, were seen with prey more often, and spent less time in the home. Generally, a wanderer was a younger cat. From the survey, the researchers found that wanderers tended to have less in the way of toys and scratching posts in the home, so they spent more time outside.

DID YOU KNOW?

Cats in the Australian study crossed an average of almost five roads per day. Unsurprisingly, given their propensity for exploration, it was the wandering cats that crossed the most roads.

Don't worry if you can't find a participatory program to join; you can record your pet's movements with GPS trackers using platforms that are readily available online. You can subscribe to services that provide you with live updates and allow you to track your cat on a computer or cell phone. There are also micro-cameras that can be fitted to your cat's collar to film his activities. This way you can find out where your cat is wandering and, of course, locate him, should he go missing.

THE FELINE PERSONALITY TEST

Do you have an adventurous or timid cat? Is she active or lazy? Impulsive or thoughtful? Is she friendly or aloof? Does she always sleep in the same spot in the house? Does she like to go out at night or prefer to stay home?

We know that cats have different personalities, just like people. The researchers working on Cat Tracker in Australia asked participants to take an online personality test on behalf of their cat before they tracked the cat's movements, based on the principle that the personality of the cat can affect her behavior when outside. Questions were asked about the cat's home environment—the provision of bedding, toys, and scratching posts—the amount of interaction with the owner, the level of access to outdoors, whether they catch prey (and if so, the type of prey), along with questions about their personality.

The Australian team analyzed the results from almost 3,000 cats, the largest-ever pet cat survey, and found that cat personalities were similar in many ways to human personalities. Psychologists have found there to be five main classifications in human personalities (neuroticism, extraversion, conscientiousness, agreeableness, and openess); the Cat Tracker results found that cats, too, have five major personality groups, which they called the "Feline Five." These are: skittishness, outgoingness, dominance, spontaneity, and friendliness. They laid their findings out in the table below.

Trait	**Characteristics**	
	Low score	**High score**
Skittishness	*Calm, trusting* The cat is well adjusted to her environment.	*Anxious, fearful* The cat may benefit from hiding spots. Something in the environment may be stressing your cat.
Outgoingness	*Aimless, quitting* Low scores are uncommon but may be linked to aging or health issues.	*Curious, active* A high-score cat may benefit from more toys and playtime.
Dominance	*Submissive, friendly to other cats* A low score may indicate the cat would adjust well to living in a multicat household.	*Bullying, aggressive to other cats* A high-score cat may have difficulties around other cats, in a multicat household and outside.
Spontaneity	*Predictable* A low score indicates the cat is well adjusted to her environment and may enjoy routine.	*Impulsive, erratic* A high-score cat may be reacting to something stressful in his environment.
Friendliness	*Solitary, irritable* A low-score cat may have a solitary nature or may be poorly socialized. If this behavior is unusual, it may indicate pain, illness, or frustration.	*Affectionate, friendly to people* The cat should fit comfortably into situations involving other cats and humans.

Among the many cats surveyed were some that had an indoor life (i.e, they were never let outside). The researchers wondered whether the personality of an indoor cat was different from cats that were allowed outside, or whether the cat's personality would change if she was kept inside. Interestingly, the indoor cats had similar personalities to those allowed to roam. So, if you do keep your cat inside, it is not having a negative effect on your pet. In fact, those kept indoors tended to be more friendly.

LEARNING NEW BEHAVIORS

As we saw in the previous chapter, a cat's ability to learn new behaviors is founded on memories and observation. Your cat will learn new behaviors by watching you. For example, he may watch you open the doors and try to attempt it himself. More imaginative cats might jump on a nearby object, such as a chair, that allows them to reach across and push the lever with their paws. Leonard Trelawny Hobhouse, a sociologist born in 1864, had a cat that learned to knock at the door. The cat didn't try to reach up and use the door knocker, but improvised by lifting up the door mat and letting it drop to the floor with a bang.

Cats can be trained into all kinds of new behaviors. The first one most owners teach when their new kitten arrives in the home is use of the litter box. I am sure you have tried to teach your cat all sorts of tricks: getting him to come when called, to sit, to fetch a toy, or sit up and beg for food, and perhaps even some agility moves, such as weaving around poles or jumping through hoops. Some people even like to get their cat to give a high five, or teach them to roll over and play dead. And, at the pinnacle of training, some cats have even been taught to play the piano and use the human toilet.

DID YOU KNOW?

Clever eye-tracking technology may be used to learn more about cat–human relationships. Employed initially to discover more about development in young children and then in dogs, it has huge potential as a means of establishing how cats look at a person after being given cues.

CLICKER-TRAINING YOUR CAT

Clicker training is a method of training through positive reinforcement. Cats quickly learn that their behavior has consequences and this method can be used to correct behavior issues. The clicker is a small plastic box containing a metal tab that, when pressed, makes a loud click sound. The idea is that, when your cat hears the click, he immediately gets a reward (usually in the form of a food treat). As a trainer, you click when the cat performs a desired behavior and you reinforce it with a treat.

Training starts with getting your cat used to hearing the click and being rewarded. The best time to try this is just before you would normally feed your cat, so he is hungry and likely to be motivated by food. Take a handful of his food and sit down with him. Take a piece of food, toss it in front of him, and immediately click. Let your cat retrieve and eat the food, then repeat the exercise. Try this a few times for a couple of days; a smart cat will soon learn what's happening and associate the click with a reward. Don't say anything, though; the click should be the clear signal. Now you can move on to the main training.

An easy exercise is to teach your cat to sit on command (shown opposite). While your cat is standing, hold a treat in your hand so the cat can see it and slowly move your hand above her head and at the same time say "sit." As the cat moves her head back to follow the treat, she will naturally lower her rear end and sit. As soon as she sits, click and treat. After some repetition, your cat will sit to the spoken command, whether rewarded or not.

USING A TARGET STICK

Some people like to use a target stick as a point of focus during training. It can be as simple as a table tennis ball attached to the end of a short piece of bamboo cane. The idea is to train your cat to look and then touch the target. You start by holding the target stick close to your cat and move it slowly in front of his head, rewarding him for any movement of his paw toward the target. From this simple beginning, you can gradually shape your cat's behavior, for example to complete a twirl. To achieve a twirl, move the target stick in front of him so his head and then his body follow the target stick, eventually completing a 360-degree spin.

TEACHING YOUR CAT TO USE THE TOILET

When I first saw a cat use a human toilet, I was seriously impressed. No more litter boxes! So how do you go about training your cat to use the toilet rather than the litter box?

1. Identify the toilet you wish them to use. Remember they need access all the time, so don't shut the door.

2. Move the litter box into the toilet and get them used to finding it there. If you are not using a flushable litter, you need to switch to one that is.

3. Training starts with raising the litter box off the ground gradually by using newspapers. Place the newspaper beside the toilet basin you wish them to use.

4. Eventually move the litter box up onto the closed lid of the toilet.

5. Once the cat is happy using the litter box on the lid, replace the litter tray with a thinnish metal tray that is strong enough to take the cat's weight, yet thin enough that you can make a hole in it (see step 7). Add litter as usual.

6. Open the lid and move the pan to inside the toilet, and make sure it is secured in place. Reduce the amount of litter you are using.

7. Make a hole in the middle and, each day, make it slightly larger. Eventually, the hole will be so large that you can remove the metal tray completely and leave the cat with the toilet basin. Don't close the lid!

The time it takes to get a cat used to using the toilet will vary. Each stage may take several days, so patience will be necessary.

The next bit of training is getting the cat to push the flush—I'll leave you to work that out!

CATS THAT LIKE WATER

While it is true that many cats hate water because their fur is not waterproof and is poorly insulated, there are some breeds that enjoy swimming or playing in water. Top of the list is the Turkish van. Its love of water has resulted in it being given the nickname "the swimming cat." Its thick coat has an unusual texture that is water repellent. The Maine coon is another breed with relatively long hair that is water resistant. It even tolerates ice-cold water. This breed has a history of being used for pest control aboard ships, so it has probably adapted as a result of being in or near water. Some owners claim that their Maine coon will even dunk their toys in their water bowls! And if you like to train your cats, the Bengal may be a good choice, because they are very good swimmers and can even be trained to fetch toys from the water.

FELINE PERFORMANCE

Moscow has a cat theater. Run by Yuri Kuklachev and his family, more than 120 cats perform alongside people and they stage a number of shows, including *The Nutcracker* and *Cats from Outer Space*. Despite Yuri Kuklachev saying that the cats cannot be trained, they perform handstands, walk tightropes, climb poles, push toy trains, and perform a variety of tricks. However, like almost all cats, they will only do something if they want to, so the performances can be quite varied. Quite often, the cats refuse to perform and either wander off or do something completely different to what was expected. The key to success, says Yuri, is to watch the cats and see how they behave naturally and then develop tricks that incorporate their natural behavior into the performance.

DID YOU KNOW?

In Japan, cats are thought to have the power to turn into super spirits when they die. This may be because, according to the Buddhist religion, the body of the cat is the temporary resting place of very spiritual people.

WHAT EMOTIONS DO CATS FEEL?

That's a difficult question to answer, because we can't put ourselves inside a cat's head. Also, we have to remember that their sensory world is very different to ours, so they are quite likely to experience feelings and emotions that we can't experience ourselves or even imagine. Nevertheless, it is possible to identify six basic emotions that all mammals experience and they are linked to finding food, play, caring and nurture, fear, anger, and social behaviors. These are basic responses that do not involve thought processes but are hardwired automatic responses that are linked to survival.

Fear: This is a primary response and a key to survival. The feeling of fear prepares the body for flight or fight, with senses heightened to assess the situation and recognize danger. A cat will run away, remain still, or hide.

Disgust: We turn our noses up at lots of things, stepping back from conditions that look unappealing or unsanitary. For cats, being disgusted by food could prevent them from eating food that is rotten or harmful.

Happiness: This emotion is linked to the release of feel-good chemicals in the brain. For a cat, a state of contentment or pleasure is linked to happiness or simply the joy of playing or hunting.

Lust: This is the urge to mate, which is seen in tomcats patrolling the streets for females that are on heat, and equally in queens when they escape the house to find a mate.

Sadness: This is seen in cats on the death of a close companion.

Anger: This is a reaction to a difficult situation and leads to the release of hormones and other chemicals to prepare the body for fight. For example, some cats may get incredibly angry when they are taken to the vets and handled against their will.

DID YOU KNOW?

Cats that suffer from behavioral problems may show extreme emotions. They may be extremely fearful, and this can manifest itself in aggression, excessive grooming, or anxiety.

In addition, there are some higher-level emotions that involve the cerebral cortex. They include euphoria, which is best seen when a cat finds catnip in the backyard, and stress, a truly complex emotion. Stress can interfere with the immune system and lead to illness. And then there is frustration, which is the result of being unable to express an emotion or agitation (imagine a cat watching a bird in the backyard through a closed window). Frustration is often manifested in the chattering of teeth. It can also be seen in cats that are not allowed out at night to hunt, and can lead to scratching or other destructive actions in the home.

Cats can pick up on human emotions, too. Do you smile or frown at your cat? Moriah Galvan and Jennifer Vonk in the USA found that cats behaved differently when their owner was smiling, compared with frowning. Cats that had a smiling owner were more likely to engage with their owner, purring and jumping on their lap. In contrast, there was less contact when the owner was frowning. When strangers replaced the owner in the test, there was no difference in the behavior between smiling and frowning, so cats had a learned response to the facial expressions of their owners, which was likely to affect the availability of a reward! So, cats really do learn to read your face!

THEORY OF MIND

This is an ability to know what another animal is thinking. If you own a dog, you probably know that when you point to something, more often than not your dog follows your gesture and understands what it means. They have learned that, by paying attention to you, they gain knowledge. American professor Brian Hare demonstrated this principle with dogs. He showed that dogs had learned that he knew where food or a toy was hidden and that, when he pointed at something, it meant he was showing them where to go find it.

Ádám Miklósi studied theory of mind with cats. He had worked with dogs and was keen to work with cats, but he soon found that there was a significant difference between cats and dogs. He produced two puzzles, one of them solvable and the other impossible. The solvable puzzle involved placing food in a bowl and securing the bowl under a stool so it was stuck. The animals had to find the bowl and pull it out to eat the food. Cats and dogs were both good at achieving this. In the impossible task, the bowl was placed under the stool, but it was tied to the stool, so that it could not be pulled out or the food retrieved. The dogs had a good go at trying to remove the bowl and then gave up. They tended to look at the person running the test as if to ask, "What should I do now?" But cats were different. They were far more determined to get at the bowl and they didn't give up so easily, nor did they look at the person for guidance.

THE FEELINGS OF OTHERS

Another aspect of "theory of mind" is picking up on the emotions of others. Again, dog owners claim that their dog is tuned into their emotional state, knowing when they are sad and giving them comfort. Brian Hare's studies, and those of other behaviorists, suggested that dogs were both self-aware and aware of the feelings of others and were displaying theory of mind. But what about cats? Cats are far more independent. They may listen to us, but, equally, they may not. And it's generally on their terms, too.

ANXIOUS CATS

Cats are surprisingly anxious pets and vets now know that cats, especially urban ones, experience more stress than first thought. It's likely that, with more cats living in close proximity in urban streets, there are more cat-to-cat encounters, causing the more nervous cat extra stress. Some confident cats will enter the homes of other cats, eat their food, and generally bully them, which causes stress. Stress is surprisingly common among cats that share a home and do not get on.

The symptoms of stress are many. The cat may groom excessively, spray urine in the house, poop outside of the litter box, be more vocal, or show aggression. It can also manifest itself in skin problems, urinary infections, and other illnesses that have a psychological element. Stress has a cause, so to solve the problem you need to work out what is causing the anxiety in your pet. It could be a change in home environment, the arrival of a baby, another cat joining the household, or bullying by a neighbor's cat, or it could be due to a change in routine, or even boredom.

DID YOU KNOW?
Licking lips can be a signal that the cat is stressed.

SEPARATION ANXIETY

Dogs suffer from separation anxiety, but do cats? We often think of cats as being independent animals that can take or leave our presence, but some people claim their cats experience separation anxiety. These owners report that, when they leave the house, their cats display unwanted behaviors, such as peeing and pooping in odd places, especially on bedding that holds the owner's smell, or they become incredibly vocal or destructive and may exhibit zealous grooming. Is this separation anxiety or simply frustration?

There has been research to indicate that some cats need their owner to feel secure and that, in their absence, the cats suffer from separation anxiety. However, more recent research from the University of Lincoln in the UK indicates that this might be simply a sign of frustration. The team observed the reactions of 20 cats when they were placed in an unfamiliar environment with either their owner or a stranger, or were left on their own. They watched each cat's responses: did she seek contact with her owner or the stranger, and did she show any signs of distress when left on her own? The results showed that the cats were a bit more vocal when they were left by their owner, but that there was no other evidence to suggest that they had a bond of attachment with their owners. In fact, the vocalization was more likely to be frustration or a learned response.

DID YOU KNOW?

I've often thought that the old adage, "curiosity killed the cat," is a bit odd. Does it mean that cats are curious and this gets them into trouble? In fact, the original phrase was, "care killed the cat," and it referred to cats' anxious nature. Indeed, cats can and do die from stress.

CHAPTER 8: TUNING INTO YOUR CAT

Meows, yowls, growls, and purrs—cats use a variety of sounds to communicate with humans and other cats. But they don't rely solely on sound to convey their emotions. Living mostly solitary lives, cats have never really needed visual signs, so their face remains relatively static, but there are subtle body movements that give insight into your cat's emotions. In fact, some signals are so subtle that we don't notice them and, even if we did, we wouldn't know what they mean.

WHAT DOES IT ALL MEAN?

Would you know that an upright tail indicates the cat is pleased to see you or that flattened ears means "leave me alone"? We may think that a cat that rolls on his back to show off his tummy wants a tickle, but, in fact, that's the last thing he wants, and you are likely to end up with his claws embedded in your hand if you try. A quick rub on the head would be more welcome. In 2013, the British charity Cats Protection surveyed more than 1,100 cat owners and identified a worrying lack of understanding of cat behavior. Two-thirds of the owners thought that purring always indicated a happy cat, while one-third didn't know that a slow blinking cat was feeling calm and contented.

DID YOU KNOW?

We don't hear sound to the same extent as a cat because they have far superior hearing. There is even a "silent meow," when a cat opens his mouth and produces a sound so high pitched that humans can't hear it. And what does it signify? We're not sure, but some experts think it's a sign of affection.

A LANGUAGE OF SOUNDS

How many sounds can your cat make? Meow, purr, growl, hiss, yowl—we can probably identify a handful, but did you know that a cat can make as many as 100 different sounds, each with a particular meaning? Cats will vary the phonetics of their sounds, too, and it is assumed that this affects the meaning of the sound, although we still have much to learn in this area.

A kitten has a relatively restricted vocabulary. They start to purr within days of birth, followed by a greeting sound of "mhm" and then a meow. As they age, they develop a much richer repertoire of sounds. Cats are quick to learn that their owner is trying to communicate and they will respond. Researchers have found that cats that have had a long-term relationship with one owner have more complex communications. Is that true of your cat?

CAT SPEAK

The meaning of cat language has long been a fascination for researchers. Back in 1944, American psychologist and cat lover Mildred Moelk produced a definitive list of 16 sound patterns made between cats and between cats and people, including hissing, shrieking, purrs, and trills, as well as six different forms of meow that she linked to the feelings of friendliness, confidence, anger, fear, pain, and annoyance. She also noted eight other sounds that she had heard vocalized when cats were fighting or mating.

CHATTY BREEDS

Some breeds are particularly talkative, especially the oriental breeds, such as the Burmese and Siamese. The Siamese has been found to have a wider vocal range, plus a more varied set of sounds. The Burmese has a narrower range, so tends to vary the length and volume of vocalizations more. In contrast, the British shorthairs are naturally much quieter, and when they do vocalize, the sounds are brief and to the point.

WHAT IS YOUR CAT SAYING?

Caterwaul: A type of yowl uttered by a female in heat, calling out to tomcats.

Chatter: Literally the chattering of teeth—this is produced usually in response to watching something exciting, such as a squirrel in the backyard, that is out of reach. There may be a chirp or a squeak, too. It is saying the cat is excited and probably frustrated.

Chirrup, trills, and chirps: These bird-like sounds can be made by a mother cat to get the attention of her kittens, and also by older cats, who may chirp to get your attention or when excited. These sounds could also be used as a greeting.

Growl: This is usually a high-pitched sound that ends with a yowl, and is made by cats that are fearful or angry. It can be a territorial sound, usually accompanied by a defensive pose.

Hiss: A sizzling sound made by a cat that is giving you, another cat, or a dog a clear warning to back off. The cat is saying that he feels threatened. It is usually accompanied by a defensive pose of raised hairs and arched back.

Meow: This is a sound used by cats when interacting with people, but never with other cats. There are different types of meows, from short mews to long meows.

Purr: This is the soft, throaty sound made by a cat.

Scream: This is a blood-curdling sound made by a female cat during mating, but which can also be made by fighting cats.

Yowl: This is a long, extended moan that is often made by a worried cat, usually to another cat. Cats yowl in many different situations: to attract a mate, to get a cat to stay away, or if the cat is unwell.

THE MEOW

A cat's plaintive meow is certain to get our attention, and it's difficult to ignore! Often, cats meow when they are hungry, but there's more than one type of meow.

Wild cats meow, but only when they are kittens. It's a call made by a kitten to gain the attention of their mother. But as soon as they leave the nest and become independent cats, they never use the sound again. Pet cats are different; they use this sound throughout their adulthood, because they have figured out that a meow gets attention.

A meow is a rising and falling sound, but there is no standard, typical meow. In fact, researchers have identified at least nineteen different meows that differ in pitch, volume, tone, and even pronunciation. A cat uses a meow to demand food, to be let out, or to get company. Some cats utter a quiet, short meow to get attention, and others a more questioning meow with a rising note, a bit like, "Can I have some food please?" But there are other forms of meow. A low-pitched meow followed by purring indicates pain, stress, or fear, while a longer meow can mean annoyance or worry, while incessant meowing may mean the cat is in pain or unwell.

DID YOU KNOW?

Whereas domesticated cats will meow to their owners for the duration of their lifetime, wild cats only do so when they are young. In the wild, the meow is something of a juvenile vocalization, used by the young to express a need. As wild cats mature, and they become more independent, the "meow" is replaced by other sounds.

LISTEN TO YOUR CAT

Try listening to your cat. How does the pattern of meows he makes vary depending on whether he is feeling happy, sad, anxious, or annoyed? Does he respond differently when you use a different style of voice? Does he prefer to be spoken to like an adult or in a singsong voice that you would use when talking to a baby?

Cats and people are similar in that both use a slightly raised tone when being friendly—for example, a cat uses a friendly chirrup. When we are annoyed or angry we use a low-pitched sound while a cat uses a low-pitched growl. Both cats and humans increase the volume to emphasize something.

DID YOU KNOW?

Some cats don't talk much because they have worked out that their owners are quite good at interpreting their body language, and so there is less of a need to vocalize.

DOES YOUR CAT HAVE AN ACCENT?

In 2016, a new five-year research project into cat dialects got underway. Based at Lund University, Sweden, Susanne Schötz and her team of researchers are comparing cat sounds from two areas in Sweden where people speak with a different dialect. They are exploring whether pet cats are influenced by the language and dialect used by their owners. The team is studying as many as 50 cats in their homes, where they are the most relaxed, to see how their mood affects the sounds they make. Do they meow when they are happy, content, hungry, and annoyed? Does a meow in Stockholm sound the same as one made by a cat living further south? As Susanne Schötz explained, cats change the melody and the intonation of their meow to send a message or show an emotion. The team is going to see if the cats respond more to "baby talk" or "adult talk." The results will help the team interpret the meaning of the sounds and enable owners to better interact with their pets.

DID YOU KNOW?

The Maine coon uses a range of chirps and trills that typically end with rising intonation, meaning that the pitch rises toward the end of the sound. Given how many people employ rising intonation when asking questions, it has been said that the Maine coon "talks in questions."

WHAT DOES A PURR MEAN?

How do you interpret a purr? Most cat owners would probably say "contented and happy." Dr. Sharon Crowell-Davis at the University of Georgia thinks differently. Her 2015 research found that owners were often misinterpreting their cat. For example, she found that purring was just as likely to be used when the cat was ill, in pain, or didn't want their owner to leave. Dr. Crowell-Davis thinks the purr can also mean "don't go anywhere."

The purr is a low-pitched sound and its volume is easily altered. Often the cat on your lap has a very quiet purr that's barely audible, and as you stroke him the purr gets louder and louder. Quite rightly, you interpret that as meaning the cat is happy. A purr can also be heard in cats that are feeling stressed or anxious; the fast, urgent-sounding purr can occur, for example, when a cat is in a strange environment or after the introduction of a new cat to the household. Your cat might also show that he is in pain by producing a continuous purr. It is thought that these purrs are linked to the release of endorphins from the brain—the body's painkillers—which are linked to healing.

CRAFTY PURRS

Cats can be crafty and often use a special purr to get their owners to do something. Pepo, a cat owned by British psychologist Karen McComb, had a knack of waking her up in the morning with insistent purring. This prompted Karen to wonder whether other cat owners had experienced similar behavior, so she investigated people's responses to different types of purring. The presence of a high-pitched cry or meow within the purr was found to be key. Food-seeking cats had higher-pitched purrs, which were more annoying and cry-like. Often, they included a high-pitched meow to make the purr more urgent-sounding and enhance the chance of a response. These purrs are known as soliciting purrs, and Karen found them used in households where there was a one-to-one relationship with the cat. There the purr was more likely to be heard and acted upon, while it was far less common in busy households, with people coming and going, and also in multicat households.

TELLING VOICES APART

Can cats distinguish between human voices? In 2013, Japanese researchers Atsuko Saito and Kazutaka Shinozuka tested cats' ability to distinguish their owners' voices from those of strangers. They recorded the voices of four strangers who were of the same gender and age as the owner. The four people were asked to call out the cat's name in the same manner as the owner. The recordings, together with the calls of the owners, were played to the cat, and the researchers looked carefully for any response, such as a twitch of the ears, a movement of the head, dilation of the pupils, movement of the tail, or a vocalization. Their results showed that cats could distinguish between the voices, with a bigger response produced when the cats heard calls from their owners. The most common response was a twitch of the ear or head movement, although some cats responded with a swish of their tail or with a sound.

BODY SIGNALS

Sounds are just one aspect of cat communication. Cats rarely open their mouth and just make a sound; as with humans and many other animals, they typically use their bodies when communicating. Their vocal communications are linked to their body language, so there are lots of signs to look for. When you bend down to stroke your cat, he may arch his back and meow, showing enjoyment of the interaction, but if he meows and pulls away it may mean the opposite, that he's had enough. If you don't pay enough attention, you can get the two behaviors muddled.

DID YOU KNOW?

When your cat rubs against your leg when you come home, he doesn't want anything, he's just giving you a cat "hug." This behavior is seen in groups of wild cats: when a cat returns, the others run up to him and they rub heads and wrap their tails around each other. It's a greeting and a way of reinforcing their individual odors among the group.

BODY TALK

Let's start with a look at the various signals that a cat may give when you meet her in the street, and what these signals mean.

I'm feeling friendly and relaxed: The cat is walking toward you with her tail up and ears pricked and slightly forward. She looks confident and she will probably give you a good sniff and may let you stroke her head and back. She's likely to rub her head and body around your legs to smear her scent everywhere.

I'm feeling very relaxed: The cat rolls on her back to expose her belly. This is the most vulnerable part of her body and she only does this if she's feeling confident, relaxed, and trusting. But this doesn't mean "tickle my belly," and an attempt to do so may end up with the cat rolling away or clawing your hand.

I'm feeling anxious:
The cat is in a crouched position with her tail tucked under or around her body. She is looking tense and her eyes are darting here and there. Don't approach her and don't cut off her escape route, but instead give her space to run away and hide.

I'm very scared:
The cat has an arched back and her fur is on end to make her look much bigger than she really is. Her pupils are dilated to the extent that she looks black-eyed and her ears are flattened against her head. She may hiss or spit if you move close. The cat is sending a very clear message that she is feeling fearful and threatened. Don't approach; leave her alone and give her space to escape.

TAIL TALK

Our cats have expressive tails. The movements can be quite subtle and complex, and they convey a lot of information, so pay attention to the tail.

Upright, vertically held

This is a friendly greeting and indicates that the cat is confident. If the tail is upright and quivering, the cat is particularly pleased to see you. He may use this tail position when asking for food. A mother cat raises her tail in this way to get her kittens to follow.

Wrapped tail

A cat that wraps his tail around your leg or arm is showing a great deal of affection.

Flicking
A slight flick shows the cat is thinking about something, but if his tail is flicking rapidly, he is anxious or agitated, especially if the tail is extended and low. A constant flicking tail shows that the cat is alert and fascinated by something in the environment.

Hooked, question mark shape
This cat is curious and interested in something, but he could be a little unsure.

Thumping
This is an annoyed or frustrated cat!

Swishing from side-to-side
Swishing can convey different messages. It could mean the cat is highly excited by something, like a bird spotted through the window, or he is enjoying something, for example when he is being groomed. Swishing can be a sign of a playful cat and he may be about to launch himself at you, but not in an aggressive way.

Between legs
This is a sign of a fearful or submissive cat that wants to be left alone.

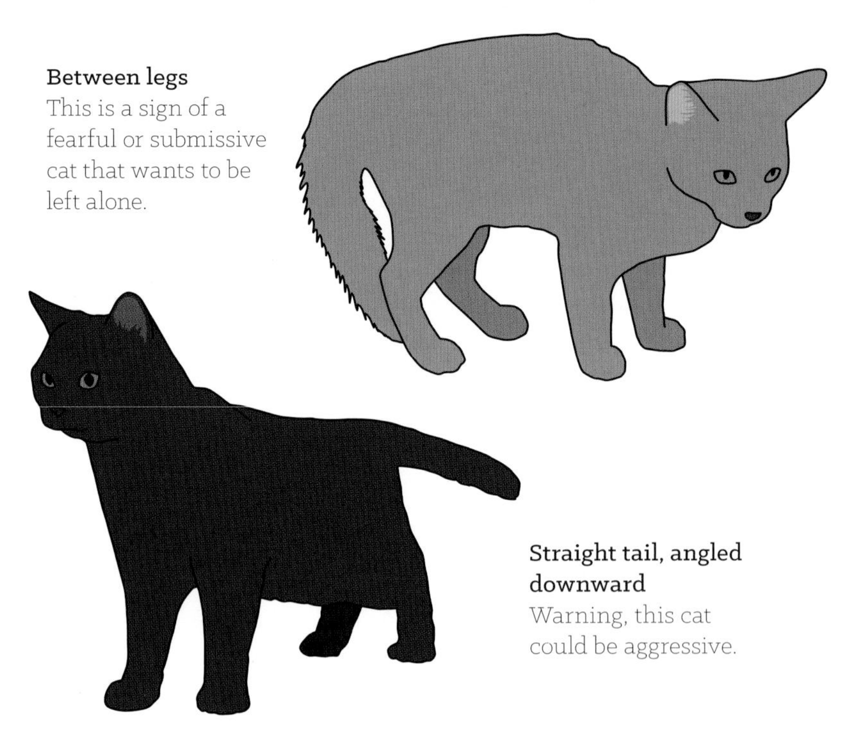

Straight tail, angled downward
Warning, this cat could be aggressive.

Straight up with hairs raised, so looking puffy
Some cats raise their tail in this way in a defensive or even angry stance, but it can also indicate a terrified cat that may be about to launch an attack.

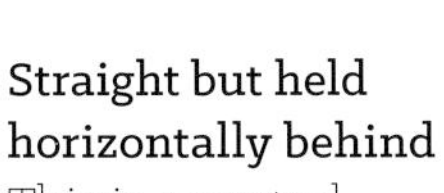

Straight but held horizontally behind
This is a neutral stance. The cat is alert, feeling relaxed and friendly.

Looping down with hook at tip
This is another defensive posture, often linked with arched back. The tail could be bristling for added effect.

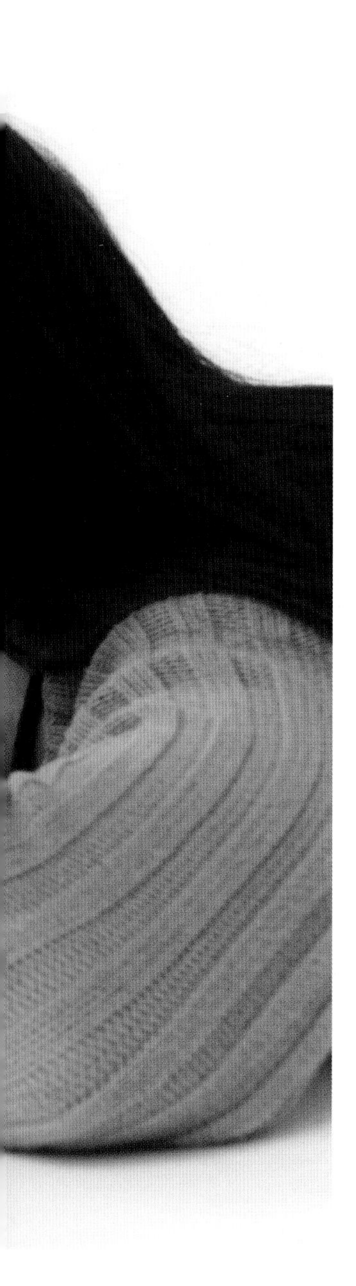

HOW MUCH DO THEY LOVE US?

We assume that we are the most important person in our cat's life and would hate to think that a complete stranger could come in and steal her affections. To prove this, in 2007, Claudia Edwards carried out the "Ainsworth's Adapted Strange Situation Test" to see if cats were more emotionally attached to their owners than to strangers. Originally, the test examined the relationship between a young child and its carer. Claudia Edward's study took 28 cats and let them experience three situations: being in a room on their own, in the same room with a stranger, and in the room with their owner. The researcher found that the cats would headbutt and play with their owner, but never with the stranger. They were more confident in the presence of their owner and were happy to explore the new surroundings, while in the other two situations they spent more time sitting by the door.

DID YOU KNOW?

The slow blink is when a cat opens and closes his eyes very slowly and turns his head from side to side. It means the cat is relaxed and not threatened.

CHAPTER 9: A SIXTH SENSE?

A cat that's always at the window when you come home, a cat running from a building just before an earthquake strikes, and another that can predict death—many people claim that cats have a sixth sense that enables them to predict things that are about to happen. Is that really possible? What is fact and what is fiction?

WAITING FOR YOUR RETURN

Is your cat always sitting on the window ledge waiting for you to come home? Many cat owners claim that their cat is waiting for them, even if they return home at different times, in different cars, and even after many months, but there is seemingly no rational explanation of this behavior.

This is an area that controversial biologist and theorist Rupert Sheldrake has investigated in cats and dogs. In 1998, his team carried out three telephone surveys in London and Greater Manchester, UK, and in Santa Cruz. In total, just under 1,000 random households were contacted and the results between the three locations were surprisingly similar. Of those people who owned cats, up to 31 percent of cat owners said their pet anticipated their return and almost half said their cats knew when they were going out before they showed any signs of doing so. About 35 percent of cat owners felt their pet was telepathic and responded to their thoughts and silent commands.

There have been various investigations into this topic. The tests have involved owners returning at random times and in strange cars, and despite this variation, more than half of the cats were at the window awaiting their return. How come? As we know, they have super hearing, so they are probably hearing sounds that we can't, such as their owner's distinctive footfall on the path or pavement. The owner may be driving a different car, but perhaps the way they change down the gears and come to a halt outside of the house is a telltale sign. And don't forget their ability to detect smells that we can't.

LIKE CLOCKWORK

Cats have a body clock, just like us. In the wild, their day is dominated by dawn and dusk, the times when they would be hunting. Our pet cats may be domesticated, but they have retained this awareness of time of day. They are creatures of habit, too. They like to do things at certain times of day, but they are also aware of our routines. Many owners report that their cat will wake them up if they sleep through their alarm or don't wake up at the usual time. Perhaps they know their owner comes back at a certain time of day, and they adapt their routine accordingly, moving to the vantage spot of a window ledge or another favored spot in readiness for their return. They may be aware of other signs that tell them the time of day; the chiming of a clock, church bells, a program on the radio or television, or the dawn chorus.

And then we must consider the selective memory of the owner. We remember the times the cat is waiting for us, but do you remember the times when the cat is not there? What happened if you were late or out for the evening? Did the cat wait for you at your usual time? Your cat may be waiting for you at the window, but how long had he been sitting there?

There seem to be plenty of rational explanations for cats waiting at a regular time, but it's impossible to explain why some cats start showing signs of their owner's return ten minutes or more before their arrival. Is it telepathy? Many cat owners say yes; they believe that their cat can read their mind.

DID YOU KNOW?

Rupert Sheldrake's exploration into the idea of telepathy between pets and their owners has proved controversial, particularly among many scientists, who question his methods. But Sheldrake's work has nevertheless generated a lot of interest among pet owners.

A TRIP TO THE VET

Talk to a vet and they say that it's not unusual for a cat owner to fail to make their appointment because the cat ran away as soon as she saw the cat carrier. Do cats pick up on our emotions? Or do they show an ability to read our minds and see our intent? Maybe it's linked to long-term memories; they remember previous occasions when they have been placed in the cat carrier and taken to a nasty-smelling place, where they were handled firmly, examined, and injected. Can you blame them for running off?

A SENSE OF DOOM

Do cats have a sense of impending doom? For thousands of years, people have claimed that cats, along with many other animals, can predict earthquakes. Records dating back to 373 BCE show that animals deserted the Greek city of Helike a few days before it was hit by a devastating earthquake. More recently, cat owners have reported that their pets acted strangely, becoming restless and agitated, and even disappearing, in the hours leading up to an earthquake.

In 1975, the sight of many animals, including cats, behaving oddly led to the evacuation of the city of Haicheng in China. An area prone to earthquakes, the officials were convinced by the animal behavior and ordered the populace to sleep outside or leave. A few days later a massive 7.3-magnitude earthquake hit the city, but thanks to the warning the number of dead and injured was much reduced. The event inevitably fueled a lot of speculation about the ability of cats and other animals to predict an earthquake. However, when the records were studied, it was noticed that the city had been hit by a number of foreshocks, so the animals were probably reacting to that. The fact that the area had suffered several other earthquakes and warnings were frequently issued was downplayed.

These reports prompted the United States Geological Survey (USGS) to conduct research into animal behavior and earthquakes. Despite investigating the claims that more pets went missing than usual in the days before an earthquake and that owners reported strange behavior, the researchers could not find anything that was significant and reliable enough to be used as an early warning system. The evidence remained merely anecdotal.

STRANGE ANIMAL BEHAVIOR

In 1942, there was a report of a sudden exodus of cats from the British city of Exeter just hours before a massive bombing raid by the Germans. Somehow the cats knew that something was going to happen. There were more reports of strange animal behavior in the hours leading up to the devastating Indian Ocean earthquake and tsunami that occurred on Boxing Day, 2004. They included reports of herds of elephants and buffalo moving to higher ground, as well as dogs and cats, and a colony of nesting flamingos abandoning their nests and flying to higher ground.

EVIDENCE FROM "ANIMAL-CAMS"

Strange animal behavior was also seen just before a strong earthquake hit Peru in 2011. A network of motion-triggered cameras was in operation around the Yanachaga-Chemillén National Park, which on a typical day recorded between five and 15 animal sightings. A team headed by Rachel Grant from Anglia Ruskin University in the UK looked at the data that had been collected in the 20 days before the earthquake struck. In the week or so before the earthquake there were five or fewer daily sightings, including five days with no animal movements at all, which was very unusual. And the rodents had disappeared completely.

Rachel Grant's team looked for reasons and found disturbances in the ionosphere above the region. The ionosphere is the layer of the Earth's atmosphere that contains a lot of ions (charged particles), which can reflect radio waves. In the buildup to an earthquake there are stresses deep in the Earth and they cause positive airborne ions to be generated at the Earth's surface. These ions cause nasty side effects in animals, which are collectively called the serotonin syndrome. When serotonin builds up in the bloodstream it causes restlessness, agitation, confusion, and anxiety—the symptoms that owners report seeing in their pets. The buildup of these ions affects ground-dwelling animals in particular, especially those living in burrows, such as the rodents that disappeared from the area.

CATS, DOGS, AND MILK YIELDS

One of the latest studies was carried out in Japan in 2011, a country that experiences more than its fair share of earthquakes. In March 2011, the country was hit by a huge earthquake and tsunami. Over the following year, Hiroyuki Yamauchi and his team carried out an online survey of pet owners. One of the questions was about unusual behavior in the days prior to the earthquake. More than 700 cat owners completed the survey along with roughly 1,200 dog owners. Around 16 percent of the cat owners and 18 percent of dog owners reported unusual behavior before the earthquake. Most reported changes just before the earthquake hit, but 30 percent of the cat owners who had noticed unusual behavior said it happened several hours before the earthquake and a few said it was up to six days before. The behaviors reported included restlessness, being noisy, meowing loudly, hiding, taking the kittens outside and simply disappearing. The researchers

knew that people may have misremembered or exaggerated the behaviors, so they studied the milk yield from cows in the earthquake areas. Farther away from the epicenter of the earthquake, the milk yields were unaltered in the lead up to the earthquake, but for those dairy herds nearest the epicenter, milk production was lower in the preceding six days.

WHAT'S THE EXPLANATION?

So how do we explain the apparent ability of cats to predict an earthquake? There are several possible explanations. We know cats are sensitive to vibrations in the ground; an earthquake is not one huge movement in the ground, but instead a number of seismic waves moving through the Earth's crust. The first is the P-wave, or pressure wave, which arrives ahead of the S-wave, the secondary, shaking wave. Animals detect the first wave and learn to associate this with danger, so they leave the house or seek shelter before the ground starts shaking.

While scientists can explain animals behaving oddly in the hours before an earthquake, giving reasons such as pressure waves, sudden rises in the static electricity, and a shift in the Earth's magnetic field, they cannot explain why an animal would behave differently days or weeks before a quake. The reason may involve subtle changes in the ground (an uplift or tilt), a change in the groundwater, vibrations from the formation of microcracks, the release of gases, the change in the ionosphere, or a small fluctuation in the electrical or magnetic fields.

EARTHQUAKE HOTLINE

Rupert Sheldrake is involved in the research into cats' responses to earthquakes. He has long believed that cats are able to predict earthquakes, and he has carried out his own research, studying the behavior of animals before major tremors, including in 1994 (Northridge, California), 1995 (Kobe, Japan), and 1999 (Greece and Turkey). He has many reports of strange behavior by pet animals before quakes, and he argues that their number and wide distribution indicate that there must be some truth in them. Sheldrake believes that a system by which people can report unusual behavior in their pets, which are then collated and analyzed, might be a means of predicting natural disasters. The incidence of false alarms and hoax calls would need to be factored in, but such a system, he claims, could feasibly be used to mitigate the effects of events such as earthquakes.

DID YOU KNOW?

During the Second World War, cats were attributed with having saved lives because they acted as early warning for approaching enemy aircraft. Their ability to detect infrasound enabled them to hear the aircraft engines well before their owners could hear them. Owners soon started to take notice of their cat and get to the safety of an air raid shelter. One such cat was Andrew, a tabby living in London. He knew when a flying bomb was going to fall near his home and took cover, as did all the people who saw him act in this way.

INCREDIBLE JOURNEYS

The book *The Incredible Journey* was a favorite among children growing up in the 1960s, and it was made into a Disney film in 1963 and again in 1993. It's the story of two dogs and a Siamese cat making a 250-mile (400-km) journey home across Canada. The story is fiction, but there are lots of real-life stories of cats making incredible journeys.

Cats have found their way to a former home after they have been moved to another house. Some have become trapped in a car or van and found themselves a long way from home. Faced with the challenge of finding their way home when confronted with unfamiliar surroundings, they made good use of their super senses to achieve the seemingly impossible.

One successful cat was Holly, a four-year-old tortoiseshell. She got separated from her owners while they were staying in Daytona Beach in Florida. Two months later, she was found a mile from the family home, having made a journey of 200 miles (320km). From her emaciated appearance, bloodied pads, and worn claws, it seemed likely that she walked home rather than having "hitched" a lift home. Holly was identified by a microchip, so she was definitely the right cat.

HOW DID HOLLY FIND HER WAY HOME?

A cat getting lost a few miles from home would rely on familiar landmarks and smells. We know cats make a mental map of their home territory—it's a bit like having an inbuilt GPS combined with the sense of smell and hearing—so it's not surprising they can find their way home over short distances, but that's not possible when they find themselves in a completely new area. Holly would probably have made use of the position of the sun and stars and even the Earth's magnetic field to find her way home. Smells can travel on the wind for many miles, so the distinctive smell of mountains, forests, and the sea might have helped.

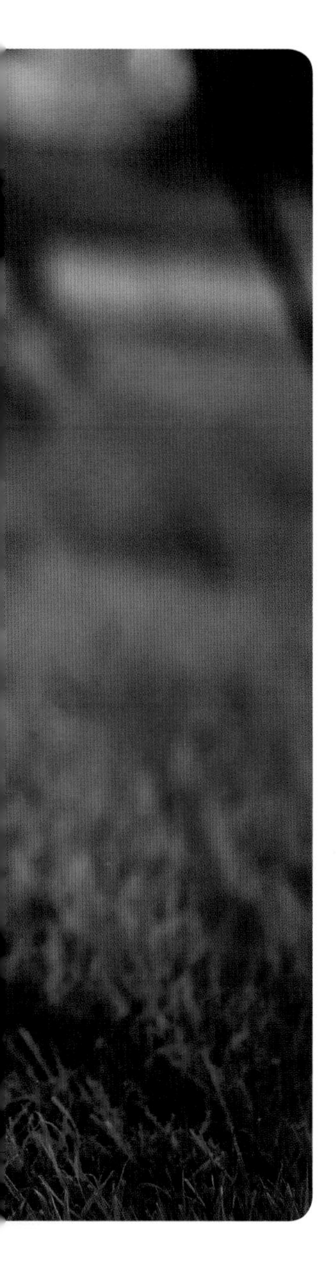

A HOMING ABILITY

There have been all sorts of experiments to investigate this homing ability, which have involved driving cats around in cars while contained in boxes, so they couldn't tell where they were going, using magnets to disorientate them, and even placing them in planetariums.

A classic experiment was carried out in 1954 by a German scientist living in the city of Kiel. He placed a number of cats in boxes, drove them around the city, and then released them into a purpose-built maze that he had constructed in a field a few miles from the city. There was a central point with 24 paths leading from it, and it was covered, so that there was no light or moonlight. Each cat was released at the central point and monitored as he explored the different paths. In most cases, the cats chose the path that led in the direction of their home, especially those living only a few miles away. They were making use of a homing ability, but nobody could explain it.

Further experiments took place in the USA, whereby the cats were drugged and then transported, so they would have no idea of where they were. Once they woke up from the sedation, they were tested in a similar type of maze and the researcher found that many picked the right route home. When the experiment was repeated with a powerful magnet near the maze, the cats were confused, showing that cats, like many other mammals and birds, had an ability to detect the Earth's magnetic field. Maybe it was this sensitivity that enabled the cats to find their way home without any visual cues.

IS SOMETHING WRONG?

With their sensitive noses, cats can detect the tiniest changes in our smell, a change that even our closest family members can't detect and one that may warn of medical problems. We know that cats can be trained as sniffer cats to find drugs and other contraband, so it's not surprising that cats can tune into our bodies and pick up changes that may tell us something is wrong.

Do a quick search on the internet and you will find numerous websites that honor pets that have had an important or lifesaving role in the lives of their owners. There are stories of cats waking up owners and preventing them from falling into a diabetic coma, having an epileptic seizure, or suffering a heart attack. The accuracy of the accounts is open to question, but the theme is consistent.

Monty, a ginger tabby, is featured in the Canadian Purina Hall of Fame. It is said that he woke up his owner by biting her fingers and prevented her from slipping into a diabetic coma. Biting the fingers was significant, because the owner used the fingers on her left hand to test for blood sugar. When she awoke she found Monty biting these same fingers—he persisted, even though she tried to shoo him off, and then she realized that she was dizzy. She staggered to the kitchen and Monty promptly sat beside her sugar testing kit. A test revealed her blood sugar to be dangerously low.

Mel-O is another hero cat. He climbed to the top bunk where a nine-year-old boy was sleeping, pawed his face, and walked on him to wake him up. The boy felt OK, but when his parents tested his blood sugar they, too, found it to be far too low. It seems Mel-O had warned them that the child was on his way to having a diabetic seizure.

Lily is also a feline medical "early warning system." It is claimed that she can detect signals that 19-year-old Nathan is about to have an epileptic fit and alerts his parents by running up and down the stairs and meowing loudly. This gives Nathan's parents time to stop Nathan hurting himself when he fits. How does she know Nathan is about to fit? It is likely that Lily picks up on tiny chemical changes that take place in the body ahead of a fit.

DETECTING CANCERS

Dogs have been trained to detect cancers. Cats are just as good at this, but because they are difficult to train and motivate, they are not used in a formal capacity. But in the home, when there is a relationship between the owner and the cat, their behavior can lead to the detection of cancers. It can manifest itself as an unusual behavior. One anecdotal tale, for example, recounts how a ginger tabby started to behave differently with his owner, climbing into his owner's bed, where he continually dragged his paw down his owner's left side. Worried, the owner visited a doctor and a lung tumor was discovered. It's not just people that cats can help, but other cats in their home. Owners report one of their cats continually licking a lump on another cat, which when investigated proved to be cancerous.

HOW DOES THIS WORK?

Why do cats behave differently around sick people? Firstly, cats can read your body language. They pick up on the subtle signs that mean something is wrong. Maybe the cat notices that their owner is holding their body slightly differently or their mood is different. They may use their super smell to detect a change in our blood sugar levels, pick up on an altered pattern of brain waves, or the presence of a hotspot caused by infection or a cancerous growth. Since cats like heat, they may choose to lie on an inflamed area because of the heat.

What about cancers? Cancers are caused by cells replicating out of control, forming tumors. The abnormal growth often produces specific chemicals that are different to the norm, and this may attract the attention of the cat, who licks or paws it, and we notice something is wrong.

INSPIRED BY PETS

Bladder cancer is one of the cancers that can now be detected early due to the development of a device called an odometer, which detects certain substances in a patient's urine. The device was inspired by the ability of cats and dogs to detect trace amounts of substances given off by the cancer. In fact, there were even trials using sniffer dogs trained to identify samples of urine that contained the substance. Now this simple test can be carried out in a doctor's office, which can lead to the early identification of bladder cancer and thereby improve survival rates.

PREDICTING DEATH

In 2007 it was reported that a cat called Oscar was predicting the deaths of people living in Steere House Nursing and Rehabilitation Center, an advanced dementia home in Rhode Island. His story, written by Dr. David Dosa, appeared in the *New England Journal of Medicine* and soon it was making headlines around the world.

Oscar, one of several cats living at Steere House, patrolled the third floor of the dementia home, visiting the rooms of all the residents. It is said he was a fairly aloof cat that was not particularly friendly and didn't like to be approached, warning people off with a hiss. The staff noticed that Oscar would only cuddle up to people in the last few days of their life. He had a natural ability to seek out those at the end of their lives, seemingly to give them comfort and company. The nursing staff soon realized that his presence on or around the bed of a patient was a clear indicator of impending death and they could warn the families accordingly. It is estimated that Oscar predicted the deaths of as many as 100 patients with almost 100 percent accuracy.

OSCAR WINS AGAIN

In one case, Oscar even proved the doctors wrong. There were two seriously ill patients, and the nursing staff was convinced that one patient was closer to death than the other, but Oscar was sitting by the other. One nurse was so worried that Oscar was wrong that she moved Oscar to the patient she believed to be nearer to death. The irate cat jumped off and returned to the other bed to resume his vigil. That patient died a few hours later, but the second patient lived for several more days.

Oscar himself had a near-death experience. He suffered a severe allergic reaction and was rushed to the vets. For a few minutes his heart stopped beating but he was revived. Fortunately, he made a full recovery and was able to return to the home and continue comforting the dying. Since Oscar hit the headlines, care homes have come forward with stories of their own "Oscars," showing that cats are helping people everywhere.

HOW DOES THIS WORK?

Clearly, people approaching the end of their life must give off a scent or a pheromone that we cannot detect but that cats can. Cells that are dying often give off sweet-smelling ketones and other substances, which could be detected by a cat with a particularly acute sense of smell.

CATS FOR THERAPY

Steere House has an ongoing animal companion program and the presence of animals, including cats, has proved to bring great benefits to the residents. Oscar was part of this program, and he was one of six cats that came from a local animal shelter. Steere House specializes in the care of late-stage dementia patients who often are unable to look after themselves. They may not be able to walk or even communicate and recognize their own family. However, they do respond to animals. Stroking a cat on their lap can help to calm them and may help to improve their mood.

Affection from cats is known to help people, and the programs have been set up that train and register animals to help people with mental, emotional, and physical problems. Cats have been found to work wonders with children with developmental disorders such as autism. The presence of a cat helps the child relax and cope with their environment, and in time become more confident. In one such program the cats are all at least one year old, with loving and sweet personalities. They must not show any sign of aggression, and should be happy to wear a harness and be petted and picked up, and they must cope well with new situations.

DID YOU KNOW?

Research shows that owning a cat has a positive impact on mental wellbeing. In 2011, UK charities Cat Protection and the Mental Health Foundation surveyed 600 cat-owning and non-owning people. Of those who owned a cat, 87 percent said the cat had a positive effect on their wellbeing and 30 percent said stroking a cat was calming and helpful. Three-quarters of the respondents said they could cope better with everyday life in the presence of cats.

CHAPTER 10: CAT PLAY

Studies on cats are relatively few and far between. Unlike dogs, cats are tricky to work with. They can be difficult to motivate, because they don't always respond to offers of treats. Equally, they can get stressed when they are taken into strange-smelling laboratories and when meeting new people. But you may have more luck when trying things out at home, so see if you can tap into your cat's intelligence and learn more about their world through simple games and exercises in your own home, where your cat is relaxed and happy to play.

Most of these tests are adaptations of the research work described in the chapters of this book. It may help to make video recordings of your cat, so you can check out the results later and look for those all-important gestures and movements that mean so much to cats. Also, when carrying out the tests, try to pick a time when your cat is in a playful mood and hungry, and therefore more likely to participate in activities in which a treat is offered as a reward.

TEST 1: CAN YOUR CAT COUNT?

This relatively simple test is about quantity discrimination. It will check whether your cat can distinguish between a larger number and a smaller number—in this case, groups of two and three black circles—and use this knowledge to find food. This is called numerical competence: telling the difference between a small number and a large number.

To complete this test you need two identical deep bowls, some treats, and two pieces of white card.

1. Take the two bowls and place some treats in one of them. Make sure the bowl is deep enough to hide the food entirely.

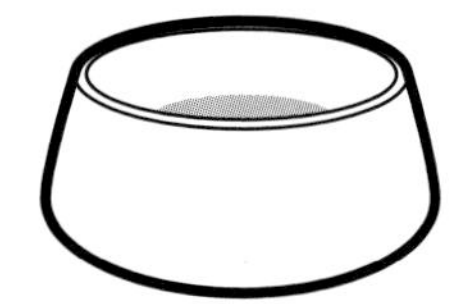

2. Take the two pieces of white card and draw two small black circles on one and three circles on the other. Make sure all five circles are the same size.

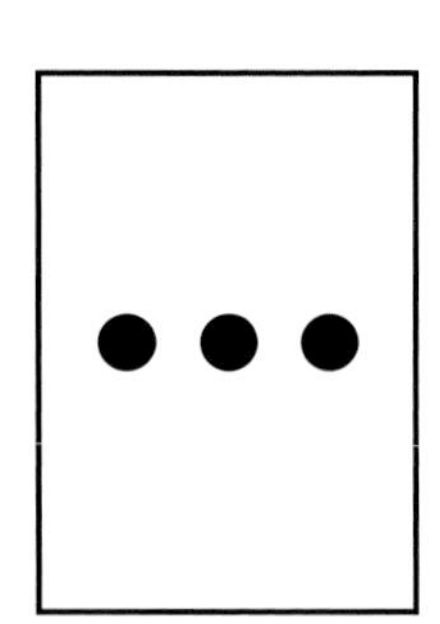

3. Place the two bowls 3 feet (1m) apart beside a white wall. Stand a card behind each of the bowls, with the two-circle card behind the bowl with treats.

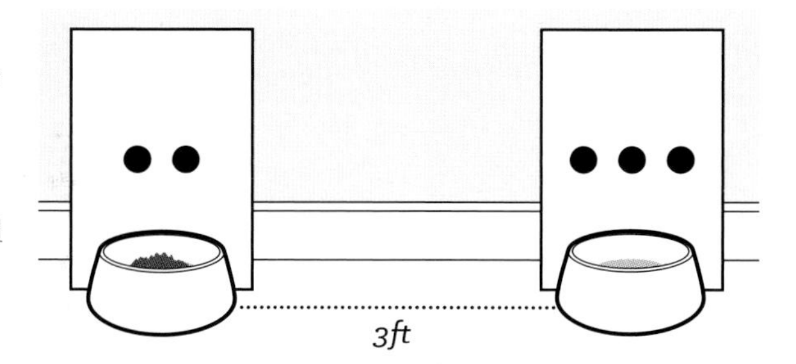

4. Let your cat into the room and allow him to investigate the bowls and find the treats. Repeat this a number of times and then repeat on the following two days. Your cat should quickly learn that two black circles means treats and run over to that bowl.

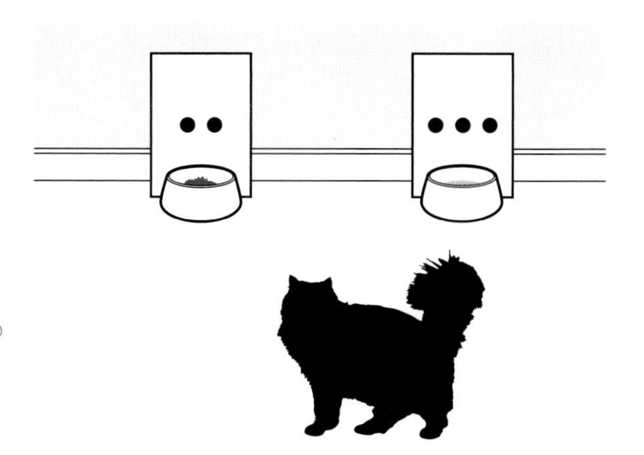

Christian Agrillo, the researcher who carried out this test, then changed the size of the circles. The two circles were made larger so that they occupied the same area as the three circles.

You can try this yourself by enlarging the size of the two circles. Repeat the test. Does this make any difference?

Consider this question: Is the size of the circle more important than the number of circles?

TEST 2: CAN YOUR CAT TELL THE TIME?

When asked whether their cat can tell the time, most owners will say yes, because their cat seems to know when they get up, and many cats are pretty good alarm clocks. They have an internal clock that tells them when it is time to wake up, to eat, and so on, but can they actually judge lengths of time? In this test you are going to attempt to train your cat to discriminate between short lengths of time. This test is simple, but it will take a few weeks to complete, because you need to pretrain your cat. Ideally, this test should be done when your cat is feeling hungry.

You will need two identical bowls, some of your cat's regular food, and some high-value treats.

1. Place the two bowls on the floor with a little food in each. Bring your cat into the room and sit down with her about 6 feet (2m) in front of the bowls. Hold the cat for a few seconds and release her. Allow her to check out the bowls. Once she has taken food from one bowl, call her back and give her a high-value treat. Repeat each day at the same time, so that by the end of the week your cat will run to either of the bowls, take the food, and come back. Pretraining is complete.

2. The next stage is to train your cat to make a choice of either the left or right bowl, depending on how long she is held for. You are aiming to train your cat to go to the left bowl after five seconds and the right one after 20 seconds. As in pretraining, you should conduct the test daily at the same time. When your cat chooses the correct bowl, give her a treat. Repeat the exercise until your cat has learned which bowl is represented by each time interval and chooses the bowl accordingly.

In the original experiment, the researcher reduced the time interval even further, so the cats had to distinguish between five and eight seconds. How did your cat do?

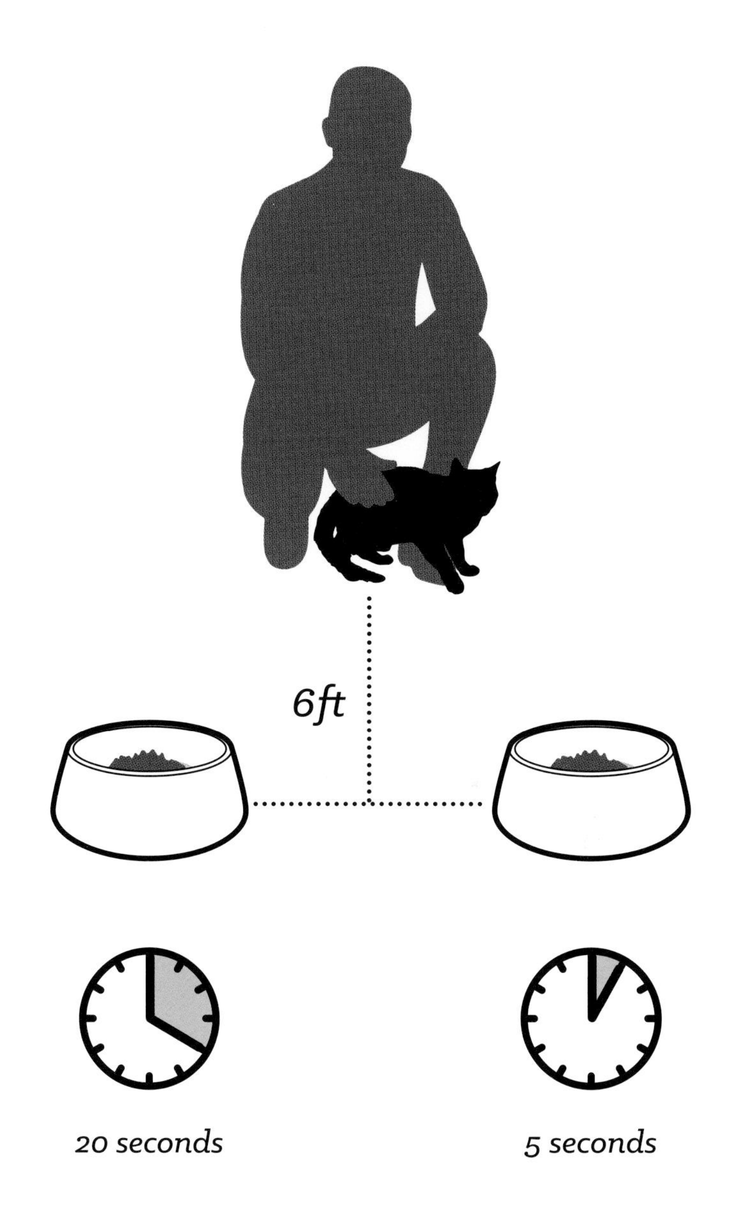
6ft
20 seconds
5 seconds

TEST 3: DOES YOUR CAT KNOW YOUR VOICE?

In 2012, Japanese researchers Atsuko Saito and Kazutaka Shinozuka proved that cats could recognize the voices of their owners. Do you think your cat recognizes your voice, or will she respond to any voice?

1. You need to record the voices of four different people—ideally, four people not known to your cat who are of the same gender and similar age as yourself. Make a recording of your own voice calling out your cat's name. Then ask each person to speak your cat's name in the same manner as you did and record it on your phone. Try to make the recordings as similar as possible.

2. You will need somebody to help you carry out the test, because you have to be out of sight. It will be useful to record a video of the proceedings, so you can watch later for subtle movements and gestures from the cat. Your helper should sit in a quiet room with your cat and play the recording of three of the strangers, one by one. The fourth voice is yours, which is followed by the voice of the fourth stranger.

3. Ask your helper to watch for any head, ear, tail, or paw movements, dilation of the pupils, or even any vocalizations. When studying the video, try to distinguish between the cat simply responding to a sound (i.e., acknowledging that a sound has been played) and showing more interest in the sound, such as pricking the ears or the pupils dilating. If she is particularly interested in the sound, she will walk toward the source of the sound.

Did your cat recognize your voice? How did she respond to the voice of the strangers?

TEST 4: KEEPING AN OBJECT IN MIND

Now we are going to venture into the world of cognitive development by carrying out an object permanence test. It's quite a complex exercise, consisting of a series of stages. It's based on tests that are commonly carried out on young children to assess their cognitive development.

The Swiss psychologist Jean Piaget developed a theory of cognitive development, explaining how a child builds a mental model of the world. A child learns about the world through trial and error, using their senses, and a cat is just the same.

Object permanence is the first stage in Jean Piaget's theory. This is the understanding that objects continue to exist, even though they are no longer visible (see page 74). There are six substages:

1. Reflex acts such as following moving objects with the eyes and closing the hand when it makes contact with an object.

2. The repeating of pleasurable acts that first happen by chance; for example, kicking the legs, wiggling fingers.

3. Ability to show more coordination and repeated actions; for example, a child shaking a rattle to hear the sounds it makes. This indicates the beginning of logic.

4. Coordination of vision and touch. Children are showing interest in objects and the things that they can do to objects; for example, they may push one object out of the way to reach another.

5. By this stage, children are exploring objects, pulling them apart and putting them back together again.

6. Children up to two years of age are able to use primitive symbols and show creativity. They are making mental representations of objects. You will be testing your cat's ability to "keep in mind" an object that has disappeared.

A child soon develops the ability to know that when a ball rolls out of sight, say under a sofa, it is still there. Cats have shown some ability in this field; for example, looking for food in places where they found it before, and continuing to hunt a mouse after it has disappeared under a shed. This proves that they have a mental representation of the object.

In this test, you will show your cat his favorite toy and then hide it from view by placing a piece of card in front of it, so that the toy is out of sight. If your cat is smart, he will look behind the card for his toy. This is an important ability. In the wild, if a cat is hunting and a mouse disappears behind a wall or under a fence, he needs to be able to remember where the mouse disappeared and even to predict where it might go while it is out of sight.

You will need a target, something that is not a treat but rather an object, such as a toy that your cat likes to play with and will want to find. It can't be too large, because it has to fit into the palm of your hand, and it must not have a smell or make a noise. You will also need a couple of plastic plant pots that you can use to hide the target and which are easy for your cat to knock over in order to retrieve the target.

You need to carry out each of the stages in turn and, if possible, on the same occasion, but your cat must progress from one stage to the next. If he fails a stage, that is the end of the test; he cannot fail a test and proceed to the next stage anyway.

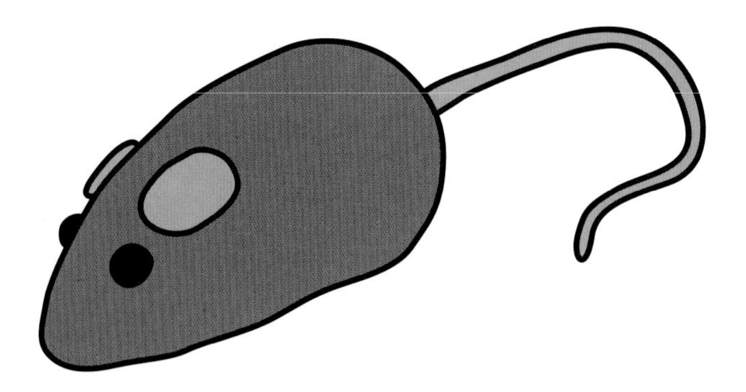

1. The first stage is to get your cat to play with the target. Try to encourage him to push the target around. If he is not cooperative at this stage, try another day, when the cat wants to play.

2. Once you have his attention, hold the target in the air, so that he has to make an effort to play with it. Hold the target at your cat's eye level and slowly move it around the head out of his field of view. Watch carefully to see if he follows the target with his eyes or moves his head. At the end of this test, let your cat play with the object as a reward.

3. Take one of the plastic pots and place it on the floor. First, let your cat sniff or even play with the pot. Then get his attention back on the target by moving the target around a bit. Once your cat is watching the target, place the target half under the pot, so that it is partly hidden. Can your cat find the target? Repeat a few times to make sure it wasn't just chance that enabled him to find the target.

4. Now you are going to hide the target completely out of sight under the pot. First, show your cat the target and move your hand slowly to the pot and place the target underneath. Make sure he is watching. Does he go up to the pot? If he does, let him paw at the pot, or even knock it over, to get at the target.

5. Now it gets tricky. You are going to increase the level of difficulty by using two pots. While your cat is watching, put the target under one pot. Repeat this several times, so that the cat is clearly going to the same pot each time. Then make a fuss of moving the target under the other pot. Has your cat followed this? Does he go to back to the first pot, or does he check out the second pot? Has he remembered where the target disappeared?

6. This is the final and highest level of achievement. Pick up the target in your hand and show it to your cat. Then close your hand and move it under the pot, carefully placing the target under the pot. It is important at this point that your cat does not see your hand open and drop the target. Remove your hand and now open it so your cat knows that the target is no longer in your hand. You can emphasize this by holding out your open hand to him. If he goes to the pot to uncover the target, he must have worked out what happened, even though he couldn't actually see the target being placed under the pot.

How did your cat do? What level did he get to?

TEST 5: PLAYING WITH TOYS

We like to give our cats toys and encourage play. Owners often notice that their cat plays with a toy as if it were a real prey animal, pouncing on it, biting it, and often pulling it to pieces. But they also get bored with their toys.

In 1992, John Bradshaw and Sarah Hall set up a study to look at cat play. The two wanted to know if cats played with their toys just for fun, or whether they saw their toys as prey. They used a selection of toys—some furry and mouselike, some feathery and birdlike, and some with lots of legs to resemble a spider. When given the toys for the first time, the cats showed a lot of interest and played with them, but when given the toys a second time there was less interest, and by the third time, the cats were clearly bored. After about five minutes, the researchers gave the cats new toys that were similar to the first ones and found that the cats' interest had been stimulated again. They felt that the cats' interest in preylike toys indicated that the cats must have some innate instincts linked to hunting and thought of their toys as real animals.

Then Bradshaw and Hall investigated the role of hunger, to see if it altered a cat's desire to play with toys. The cat's first meal of the day was delayed and she was given a mouselike toy instead. The researchers predicted that a hungry cat wouldn't be interested in playing because her mind would be on hunting and finding food. But the result was just the opposite; the hungry cat set upon her toys with gusto, proving to the researchers that cats are thinking about hunting when they play with toys.

You can try this experiment with your own cat. You will need two toys that are animal-like, such as furry mouse and bird toys.

1. Give your cat one of the animal-like toys. Does she play with it? If she's not interested in play, take the toy away and try again on another occasion.

2. After a few mintues' play, take the toy away. Wait five minutes and give the toy back. Repeat this a couple of times. Does your cat remain interested in the toy, or does she get bored?

3. Wait five minutes and give your cat the other animal toy. Is she more or less interested in the toy?

In the second part of this test, you are going to see whether hunger affects the way your cat plays with the toy.

4. If you feed your cat in the morning, try this test before you feed her. Instead of feeding your cat, give her one of the toys. Is she interested in the toy? Does she play or does she meow for her food?

TEST 6: CAN YOUR CAT SOLVE PROBLEMS?

How often have you seen your cat use his paw to pull things closer, or to retrieve a treat stuck under the sofa or something floating on a puddle?

In this test you are going to gauge your cat's intelligence to see if he can solve a problem. He has to work out whether there is a connection between seeing a treat attached to one end of a string and pulling on the other end to bring it closer. The test builds up from using one string, then two strings, and finally two crossed-over strings.

Your will need some string, smelly treats, and a plastic screen.

1. Cut two 20-inch (50-cm) lengths of string.

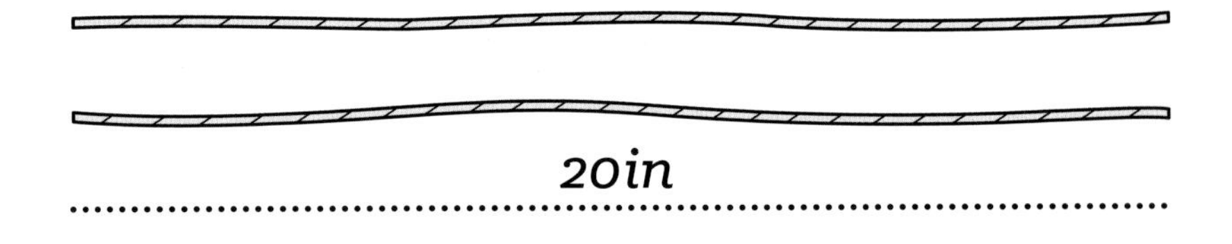

2. Attach the treat to the end of one piece of string.

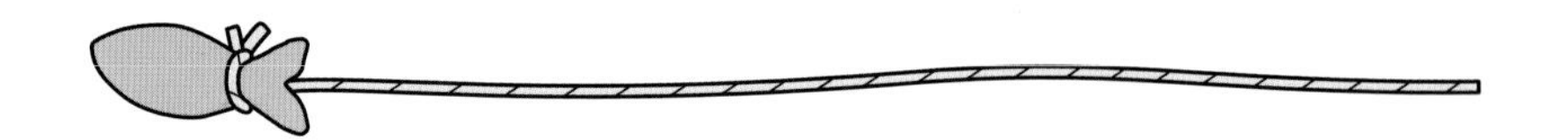

3. Now lay out the length of string on the floor and place the plastic screen on top, so that the treat is in the middle and the other end is sticking out of the screen. The screen needs to be slightly off the ground, so that when the cat pulls the string, the treat will move and not snag on the screen.

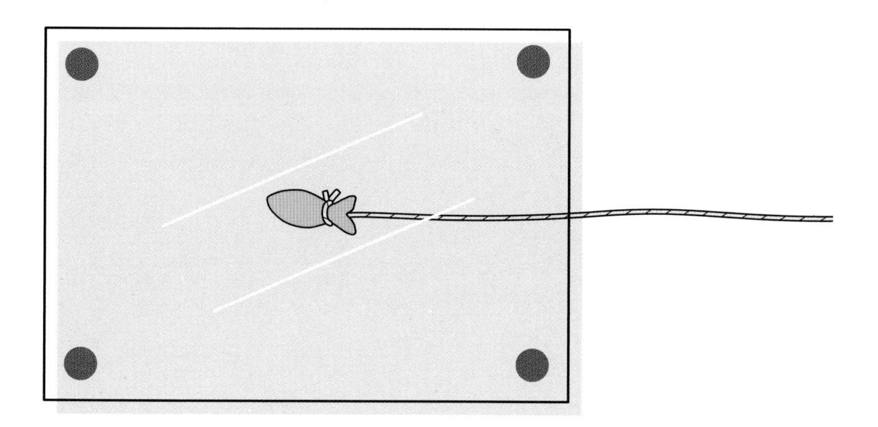

4. Bring your cat in the room and release him near the screen and the free end of the string.

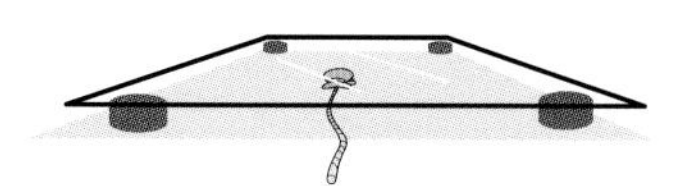

5. Watch to see if your cat pulls on the string to drag the treat.

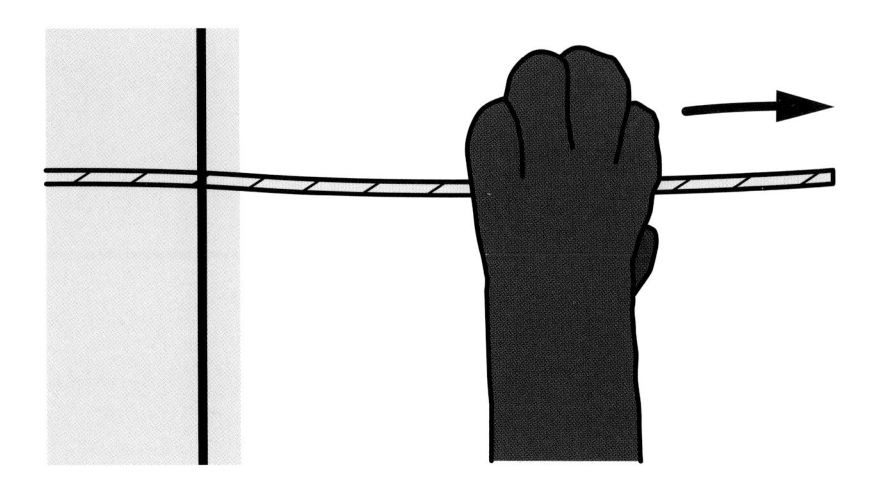

6. If your cat is successful and retrieves the treat, build up the difficulty by placing two pieces of string parallel under the plastic screen, with the treat attached to one. Does your cat work out that the treat is only attached to one string?

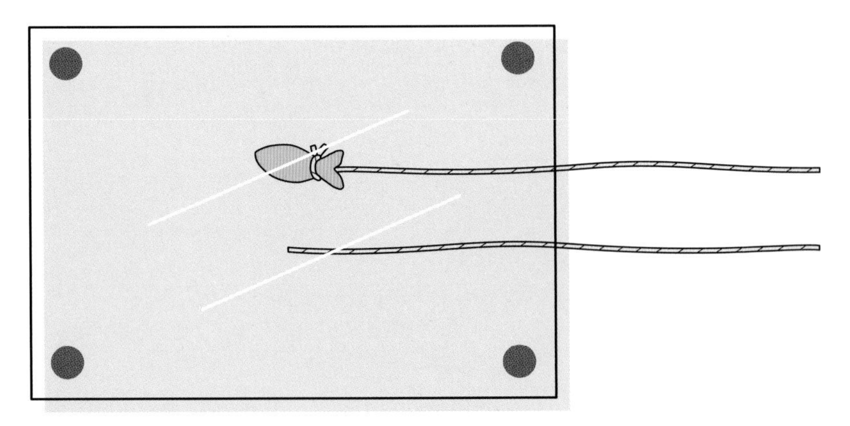

7. Finally, to make it more difficult still, cross over the strings and see if your cat can work out which string to pull to retrieve the treat. This is difficult and few animals succeed at this level, so well done if your cat can work this out.

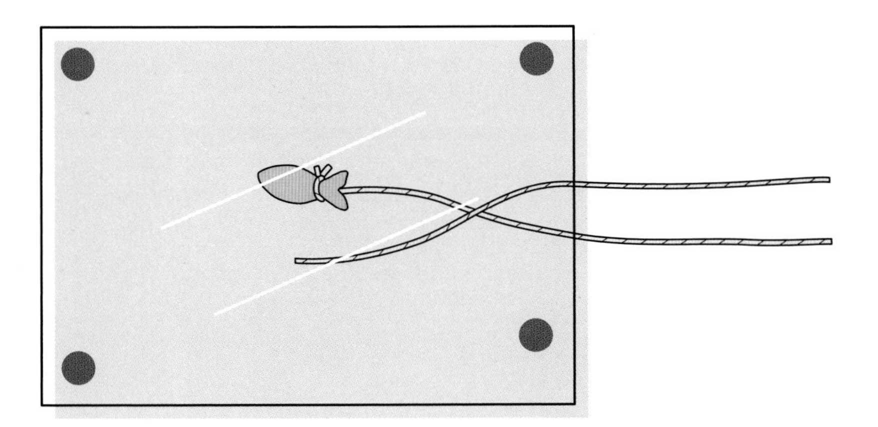

TEST 7: CAN YOUR CAT UNDERSTAND YOUR GESTURES?

This test investigates the idea called "theory of mind." Chapter 7 (page 105) describes the research by Hungarian researcher Ádám Miklósi, and in this test we are going to recreate the experiment that he carried out with cats in a home environment. Be warned, he didn't find this very easy, so you may need to persevere!

For this test, you will need two identical brown plastic pots approximately 5½ inches (14cm) in diameter and 4 inches (10cm) high, some small high-value food treats, and somebody to help you.

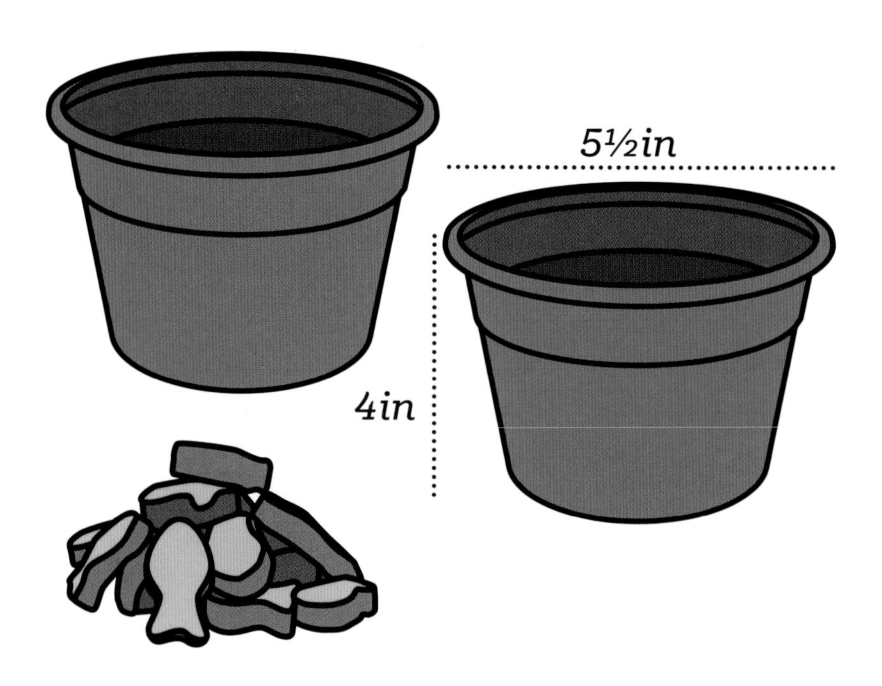

1. In a quiet room, sit on the floor and hold your cat. Ask your helper to place the two pots 5 feet (1.5m) apart on the floor and about 8 feet (2.5m) in front of you.

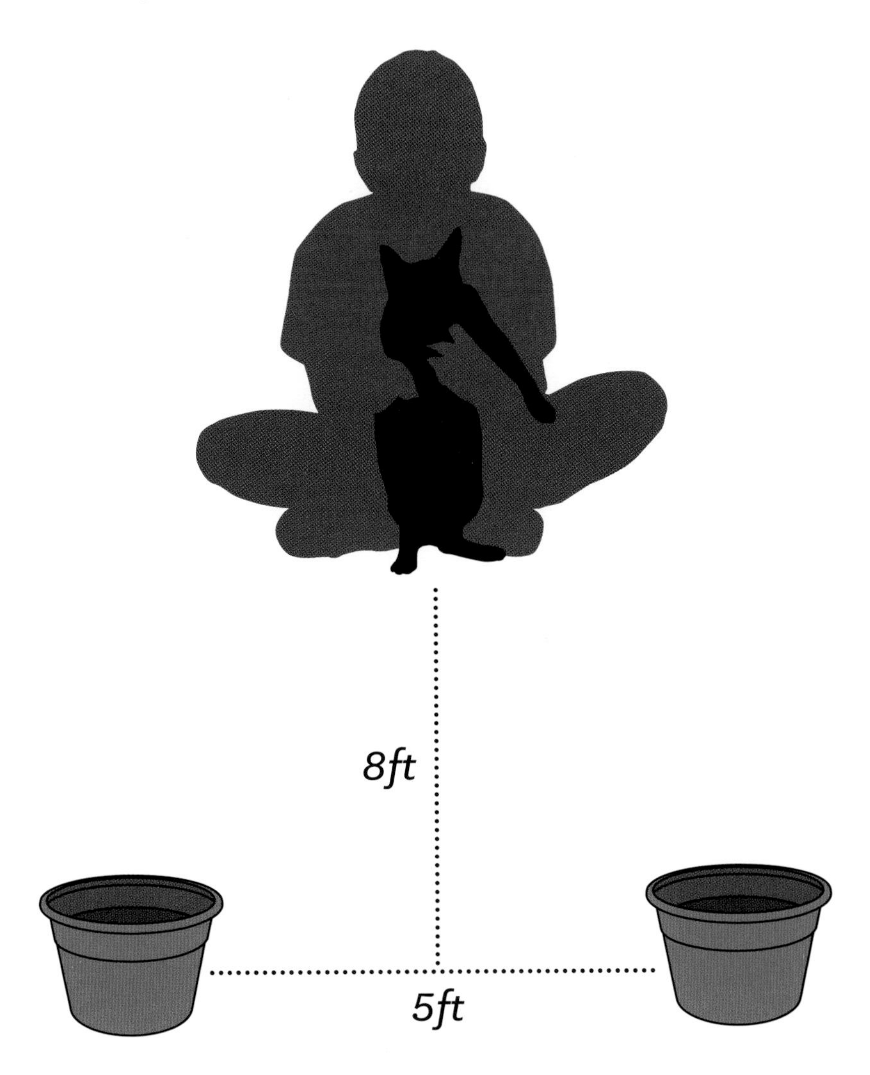

2. While your cat is watching, the helper places a treat in one of the pots. Release the cat and let her retrieve and eat the treat.

3. Repeat this twice more for each pot, so that your cat knows that both pots may contain food.

4. Now the process is repeated, but this time your cat is prevented from seeing the treats being placed in the pot. The helper puts a treat in one pot without your cat seeing and then places them on the floor as before. With your cat in position, the helper sits behind the pots in a central position, equidistant between the two pots.

5. First your helper gets your cat's attention by clapping or calling her name, and then points at the pot with the treat. If your cat does not respond, the process of calling and pointing is repeated. Your cat is only allowed to choose one pot, and should not be allowed to go from the incorrect pot to the correct one. This is repeated three times.

At all stages, if your cat loses interest or walks away, stop the test and try again when she is more playful or interested.

Does your cat understand the arm gestures made by your helper and go to the correct pot?

TEST 8: IS YOUR CAT RIGHT- OR LEFT-HANDED?

We all know that most people are either right- or left-handed, but what about your cat? Research conducted in 2009 by Deborah L. Wells and Sarah Millsopp at the Animal Behaviour Centre, Queens University, Northern Ireland found a preference in cats, but surprisingly it was linked to their sex. They found that female cats preferred to use their right paw when carrying out a task, such as removing food from a jar, while male cats preferred their left paw. What about your cat? Here are some simple tasks to test your cat's preference.

The researchers carried out three simple tasks, which you could try with your cat. For each of these three tasks, record which paw was used.

Task 1: Place a treat in a large, wide-necked jar. Let your cat explore and try to remove the food treat. Which paw is used?

Task 2: Suspend a toy or treat above your cat. Which paw did he extend to reach the toy?

Task 3: Pull a toy along the ground and let your cat try to catch it. Which paw did he use to grab the toy?

Repeat these tests once per day, and at the end of the week see if a pattern emerges.

TEST 9: CAN YOUR CAT SOLVE A FOOD PUZZLE?

Do you put your cat's food out in the same bowl and in the same place every day? Have you thought of changing it? Scientists think that giving your cat more of a challenge will leave him healthier and happier, and it can give a cat with behavioral problems a positive boost.

The idea behind this test is to recreate the way cats hunt for food in the wild. First, they have to find the food and then work out how to reach it.

The test is based on a food puzzle, which is ideal if you feed your cat dry kibble. If you feed a wet or raw diet, use high-value treats. You need a container that can hold the dry food and which is easy for the cat to move around, something like a plastic jar or a see-through ball in which you can make small holes. As the cat plays with the container, the food drops out of the holes.

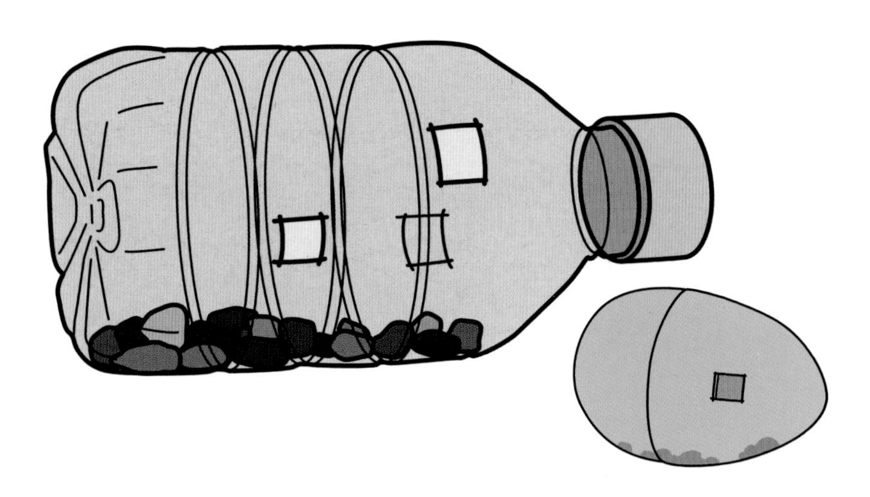

1. The first stage is to put the container down in the place where you would normally feed your cat.

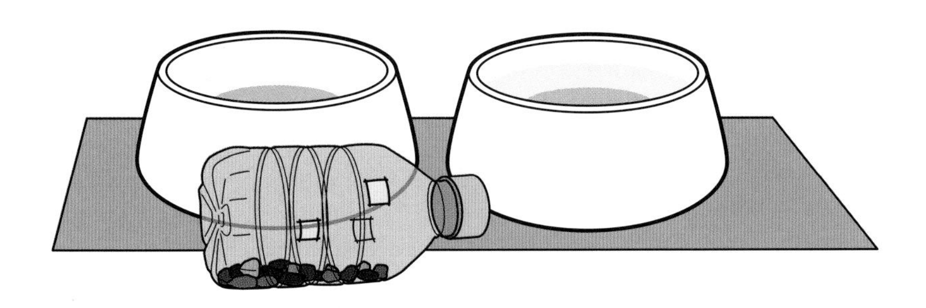

2. Let your cat play with the puzzle and work out how to get the food to drop out.

3. Once your cat knows that his food is inside and knows how to get the food to fall out, start to make the test more difficult. This can be done by covering over some of the holes, so that there are fewer holes and the cat has to work harder to get the food to fall out.

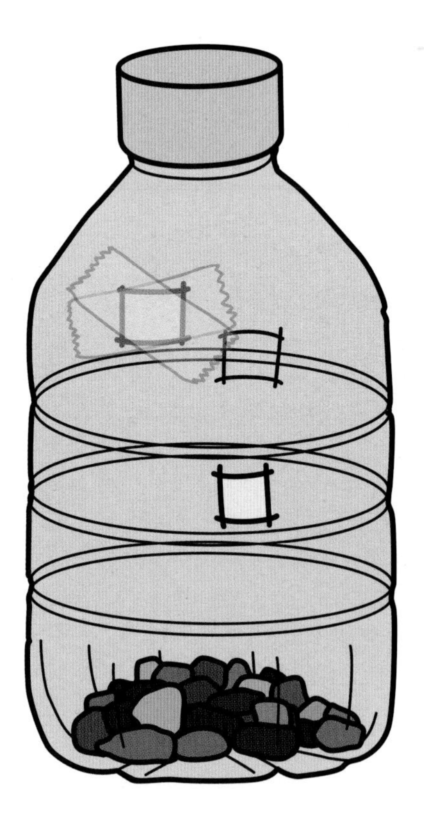

4. You can increase the level of difficulty again by changing the position of the container.

5. Another variation is to switch to an opaque container, or to use a shape that is less easy to move.

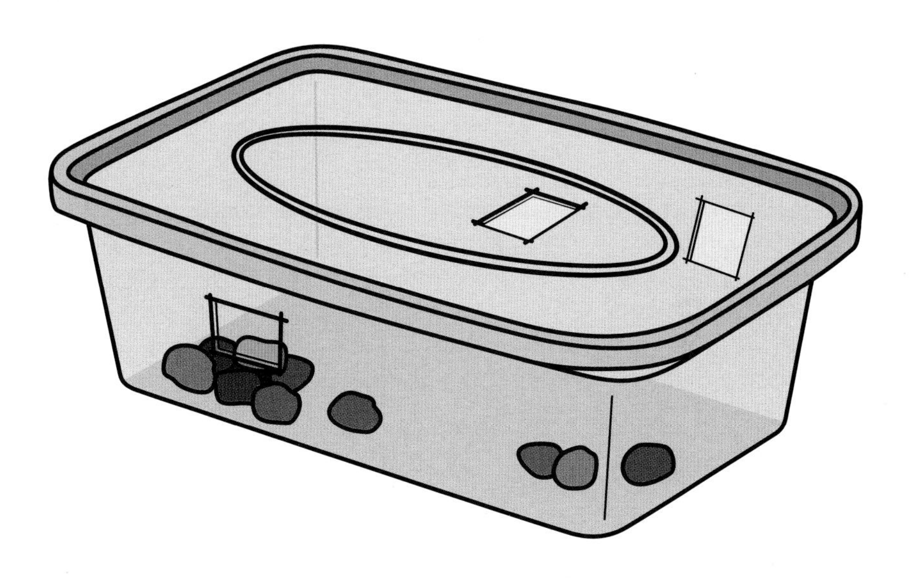

6. Finally, you could make it even more difficult by hiding the container, so that your cat has to hunt for it and then get the food out.

GLOSSARY

Acuity: The ability of the eye to see detail

Amplify: To make louder

Amplitude: The size of a sound wave

Cartilage: A strong and flexible material that supports and protects joints

Cerebellum: The part of the brain involved with movement, posture, and balance

Cerebrum: A large part of the brain, which is highly folded and involved with behavior, emotions, and memories

Cognitive dysfunction: The loss of brain cells and the disruption of memories, which can lead to dementia

Cone: A receptor cell in the retina of the eye responsible for color detection

Decibel: The unit by which the loudness, or intensity, of sound is measured

Flehmen response: A behavior seen in cats that allows them to analyze a smell

Frequency: The rate at which something occurs in a specific period of time

Gene: The unit of heredity

Habituation: Repeated exposure to a stimulus

Innate: Natural; inbuilt

Ionosphere: A layer of Earth's atmosphere

Magnetic field: The field of force around a magnet, which is also present around Earth's core

Mutation: A sudden change in DNA that may alter the appearance of the individual

Neurone: A nerve cell

Nictitating membrane: A third eyelid

Object permanence: The ability to keep something in mind when it disappears from sight

Pheromones: Chemicals produced by animals

Phonetics: The sounds of speech

Pitch: The frequency of a soundwave

Pupil: The circular opening in the center of the iris that lets light through

Receptor: A cell or group of cells able to detect external stimuli, such as taste and sound

Retina: The light-sensitive layer at the back of the eye

Rod: A receptor cell in the retina of the eye enabling objects to be seen in low-light conditions

Seismic: Vibrations or tremors in the ground

Serotonin: A chemical found in the body that is involved with nerve cells, pain reception, and the sleep cycle

Tapetum: The reflective layer at the back of a cat's eye

Taste bud: A group of receptors on the tongue that detect particular tastes

Ultrasonic: High-pitched sounds that humans cannot hear

Ultraviolet: A form of light radiation that humans cannot see; it lies beyond the violet end of the visible spectrum

Vibration: The rapid backward and forward movement of an object

Wavelength: The distance between two troughs or two crests of sound waves and other waveforms

FURTHER RESOURCES

Bradshaw, John. ***The Behavior of the Domestic Cat, second edition***.
Oxfordshire: CABI Publishing, 2012.

Bradshaw, John. ***Cat Sense: The Feline Enigma Revealed***.
London: Penguin, 2014.

Morris , Desmond. ***Catwatching: The Essential Guide to Cat Behavior***.
New York: Ebury Press, 2002.

Roberts, Dr. Gordon. ***Understanding Cat Behavior***.
CreateSpace Independent Publishing Platform, 2014.

Turner, Dennis. ***The Domestic Cat: The Biology of its Behavior***.
Cambridge: CUP, 2013.

INDEX

CREDITS

7: © Irina Fischer

9: © Cressida Studio

10: © Deep OV

13: © Evgeny Eremeev

15: © Anna Bolotnikova

16 Top: © Africa Studio

16 Bottom: © Africa Studio

17: © Olesya Tseytlin

18–19: © Torie McMillan

20 Top: © Lilia Beck

20 Middle: © VladJ55

20 Bottom: © Drawen

23: © DenisNata

27: © Anna Hoychuk

28: © Schankz

29: © Agata Kowalczyk

31 Top left: © Stefan Petru Andronache

31 Top right: © Cherry-Merry

31 Bottom: © 5 Second Studio

33: © Tocak

35: © Hemerocallis

36: © Eric Isselee

37: © MAErtek

39: © StockPhotosArt

40: © Zandyz

41: © Maximult

43: © Alta Oosthuizen

45: © Julie Src

46–47: © Gvictoria

49 Left: © Joanna Zaleska

49 Right: © Anna Sedneva

50: © GooDween123

52–53: © April Turner

55: © Fantom_rd

56 Top: © Pavel Sazonov

56 Bottom: © Art_man

60: © Maradon 333

62–63: © Patrick Lienin

67: © Pascale Gueret

68: © Pavel Litvinsky

71: © Kosikhina Anna

73: © Seregraff

74: © Quang Nguyen Vinh

77: © Anurak Pongpatimet

78: © Nina Buday

80: © Alexx60

83: © Orhan Cam

85: © Vvvita

86: © MaraZe

89: © Paul W. Thompson

91: © Khamidulin Sergey

93: © 5 Second Studio

94–95: © Linavita

98: © Daria Berdnikova

101: © Andrey Khusnutdinov

103: © Hannadarzy

104: © Smolina Marianna

106–107: © Trybex

108: © Africa Studio

111: © MiQ

112: © Dagmar Hijmans

115: © Hannadarzy

116: © Robert Petrovic

119: © eZeePics

120: © Luna Vandoorne

123: © Gordana Sermek

124 Top: © Eric Isselee

124 Bottom: © Susan Schmitz

125 Top: © Susana Reyes

125 Bottom: © De Jongh Photography

130–131: © Kitty

133: © lkoimages

135: © Foonia

136: © Alena Stalmashonak

139: © Sari O'Neal

140: © KPG_Payless

143: © Natalia Fadosova

144–145: © Stephen Moehle

147: © Supanee Sukanakintr

148: © Graphbottles

151: © Budimir Jevtic

152: © Budimir Jevtic

154: © Andrey Kuzmin

155: © Benjamin Simeneta

163: © Ysbrand Cosijn

188: © Tsekhmister